The Future Use of Nordic Forests

Erik Westholm • Karin Beland Lindahl
Florian Kraxner
Editors

The Future Use of Nordic Forests

A Global Perspective

Springer

Editors
Erik Westholm
Department of Urban and Rural Development
Swedish University of Agricultural Sciences (SLU)
Uppsala, Sweden

Karin Beland Lindahl
Unit of Political Science
Luleå University of Technology (LTU)
Luleå, Sweden

Florian Kraxner
Ecosystems Services and Management Program (ESM)
International Institute for Applied Systems Analysis (IIASA)
Laxenburg, Austria

This title will be printed globally on one of the Chains of Custody certified paper stock, including but not limited to FSC, PEFC and SFI.

ISBN 978-3-319-14217-3 ISBN 978-3-319-14218-0 (eBook)
DOI 10.1007/978-3-319-14218-0

Library of Congress Control Number: 2015931516

Springer Cham Heidelberg New York Dordrecht London

Selection of photos (cover and chapter illustrations) by Eva Kvarnström

Printed on acid-free paper

Springer International Publishing AG Switzerland is part of Springer Science+Business Media (www.springer.com)

Foreword

The vast boreal forests have played a key role in the establishment of the Nordic welfare states. Sawn timber, pulp, and paper are major export products, and Nordic forest companies have a long-standing record of operating on the world market. While forestry and forest industries provide thousands of rural jobs, the forests also deliver other ecosystem services, and recreational and amenity values.

The debate on how to strike a balance between the various functions of the forests must always be scientifically grounded. *Future forests* is a strategic research program addressing the future of the Swedish forests. It operates in collaboration between the Swedish University for Agricultural Sciences, Umeå University, and Skogforsk, and is funded by a group of public and corporate organizations covering key forest interests in Sweden. The researchers in the program represent many disciplines and their expertise embraces the broad field of forest-related issues, from the biological functions of forest environments to the political issues pertaining to the multiple use of forested land in Nordic countries.

This book presents results from one of the component projects: *Forestry at the cross-roads*. The project has explored global trends that may affect the future use of boreal forests. The growing world population and the expected economic growth in countries with huge populations, the urgent need for solutions to the energy crisis, and the worrying climate scenarios are challenges that represent a quite different future, including that of boreal forests. Together, they suggest a growth in critical demand, indicating a pressure on all land-based resources. The future is always filled with uncertainties; however, it seems inevitable that these expected developments will and must impact long-term strategic decisions on the future use of Nordic forests.

The authors of this book represent a distinguished group of internationally renowned scientists from various academic fields covering political science, geography, rural development, forest economics, history, and geo-sciences.

It is hoped that the book will interest a broad range of readers, including academics, strategic actors in the sector, policymakers, and also a general public interested in natural resource futures.

Maria Norrfalk
Executive committee chairman: Future Forests
Governor of County Dalarna, Sweden

Acknowledgments

We are deeply grateful to the authors who contributed chapters to this unique book project on the Future Use of Nordic Forests. The authors' own academic institutions played an important role in supporting their work and guiding our efforts overall.

Special thanks go to the large team of renowned external experts who carried out a double-blind peer-review process of the book and helped us achieve the highest possible scientific quality.

Thankfully, we acknowledge the dedicated efforts of Maria Norrfalk, Executive Committee Chairman of Future Forests and Governor of Sweden's County Dalarna; and Alexander Buck, Executive Director of the International Union of Forest Research Organizations – IUFRO, in providing the foreword and the endorsement on the back cover of this book.

We express our gratitude to the editors' institutions and departments, their excellent support staff, and the background funding projects. These are the Department of Urban and Rural Development of the Swedish University for Agricultural Sciences (SLU), the Political Science Unit of the Luleå University of Technology (LTU), and the Ecosystems Services Program (ESM) of the International Institute for Applied Systems Analysis (IIASA).

Last but not least, we thank both the Future Forest and the Future Agriculture programs of Sweden.

The research was funded through Future Forests, a multi-disciplinary research program supported by the Foundation for Strategic Environmental Research (MISTRA), the Swedish Forestry Industry, the Swedish University of Agricultural Sciences (SLU), Umeå University, and the Forestry Research Institute of Sweden.

Contents

About the Editors and Authors

Erik Westholm (editor)

Professor, Department for Urban and Rural Development, Swedish University of Agricultural Sciences (SLU), Uppsala, Sweden.

E-mail: erik.westholm@slu.se

Erik Westholm is Professor in Social and Economic Geography at Högskolan Dalarna and in Nordic and European Rural Development at the Swedish University for Agricultural Sciences, SLU. His research interests cover the future of rural areas; social and economic change as well as the future use of Nordic forests and farmed land. He also has an interest in the future as a research field; perspectives, methodologies and genres in futures studies. He is the Director of the strategic platform, Future Agriculture and was also project manager within the Future Forest program. He is Chair of the Scientific Advisory Board of the Swedish Agency for Nature Conservation (Naturvårdsverket). Westholm is a Member of the Royal Swedish Academy for Agriculture and Forestry, KSLA.

Karin Beland Lindahl (editor)

Assistant Professor, Unit of Political Science, Luleå University of Technology, Luleå, Sweden.

E-mail: karin.beland.lindahl@ltu.se

Karin Beland Lindahl is Assistant Professor with the Political Science Unit of Luleå University of Technology. Her research interest is in the politics of natural resource management, especially that of forests, minerals and renewable energy. She uses policy analysis, particularly interpretive approaches, to explore actors' understandings of policy and natural resource conflicts. Her work ranges from analyses of local controversies to studies of global forest trends. Dr. Beland Lindahl is affiliated with the Future Forests program, currently analyzing how different forest governance models respond to various sustainability challenges. She is also engaged in a project exploring controversies over mining and pathways to sustainability in the Swedish mountain region.

Florian Kraxner (editor)

Deputy Director, Ecosystems Services and Management (ESM) Program, International Institute for Applied Systems Analysis (IIASA), Ecosystems Services and Management Program (ESM), Laxenburg, Austria.

E-mail: kraxner@iiasa.ac.at

Florian Kraxner is Deputy Director of the Ecosystems Services and Management (ESM) Program at the International Institute for Applied Systems Analysis (IIASA). He was previously Deputy Leader of IIASA's Forestry Program for which he has been conducting studies in the field of integrated land use -modeling. He is a co-author with IPBES and also serving on various scientific boards within the research areas of Future Earth – with special emphasis on sustainability. On behalf of his institute he has been appointed the official representative to the International Environmental Science Initiative SCOPE and the Coordinator of the UK Policy-Science Activity FORESIGHT. Furthermore, he leads ESM's Research Group on Policy-Science Interface (PSI) where he and his team are currently focusing on the socioeconomics and policies of the LULUCF sector, forest-based bioenergy, renewable energy systems optimization, ecosystems services integration, and the linking of bioenergy with carbon capture and storage (BECCS).

Katarina Eckerberg

Professor, Department of Political Science, Umeå University, Sweden.

E-mail: katarina.eckerberg@pol.umu.se

Katarina Eckerberg is Professor in Public Administration at the Department of Political Science, Umeå University, Sweden. Her research comprises the field of policy analysis and implementation studies of environment and sustainable development policy from international to local level, often with a European comparative perspective, including studies of land use systems, environmental policy instruments, and Local Agenda 21. Currently, she is engaged in research on collaborative governance in the Swedish mountain region, water management and ecological restoration in policy and practice. With over 130 publications to date she has served on a large number of research advisory boards, been community forestry officer at the FAO in Rome and Deputy Director for Stockholm Environment Institute.

Bo Malmberg

Professor, Department of Human Geography, Stockholm University, Stockholm, Sweden.

E-mail: bo.malmberg@humangeo.su.se

Bo Malmberg is Professor in Human Geography at Stockholm University. His main research interest is demographic effects on social and economic development at the regional, national, and global level. From 2000 to 2013 he was affiliated with the Institute for Futures in Stockholm where, in collaboration with Professor Thomas Lindh, he developed tools for demographically based long-term projections of economic trends. His work in this field includes studies for the European Commission, the Swedish ministries of finance and the environment, the Swedish Trade Union Federation, Riksbyggen, and MISTRA. He has also developed tools for analyzing land use systems and urban structure using remote sensing data, and improved spatial measures of demographic composition.

Sten Nilsson

Member, The Royal Swedish Academy of Agriculture and Forestry.

Guest Scholar, Ecosystems Services and Management Program (ESM), International Institute for Applied Systems Analysis (IIASA), Laxenburg, Austria

Email: stenbnilsson@gmail.com; nilsson@iiasa.ac.at

Sten Nilsson is a Professor in economic planning, has had a distinguished academic career in forest sector analysis with the emphasis on policy analysis. He was initiator and leader of the Institute of Forests, Industry and Market Studies of the Swedish University for Agricultural Sciences. Professor Nilsson joined the International Institute for Applied Systems Analysis in 1986, where he was Deputy and then Acting Director from 2002 to 2009. He is currently also serving as independent adviser to governments and industry on transformation of the forest sector. He is a Guest Scholar, Ecosystems Services and Management (ESM) Program, IIASA, and Member, The Royal Swedish Academy of Agriculture and Forestry. He has held a number of consultancies in industry, governments, and organizations such as The World Bank, FAO, OECD, European Commission, and SIDA.

Eva-Maria Nordström

Researcher, Swedish University of Agricultural Sciences (SLU), Department of Forest Resource Management, Umeå, Sweden

Guest Scholar, Ecosystems Services and Management Program (ESM), International Institute for Applied Systems Analysis (IIASA), Laxenburg, Austria

E-mail: Eva-Maria.Nordstrom@slu.se; nordstr@iiasa.ac.at

Eva-Maria Nordström holds a PhD in Forestry and works at the Department of Forest Resource Management at the Swedish University of Agricultural Sciences (SLU). Her research focuses primarily on forest management planning and policy situations with conflicting values and multiple stakeholders. She has carried out her postdoctoral studies at the Ecosystems Services and Management (ESM) Program at the International Institute for Applied Systems Analysis (IIASA), in Laxenburg, Austria, during 2012–2013. At IIASA, she has been working on global and local scenario analysis for the forest sector with a special focus on Scandinavia and Sweden. Dr. Nordström is currently coordinating the collaboration between IIASA and the Future Forest Programme on this topic.

Camilla Sandström

Senior Lecturer and Associate Professor, Department of Political Science, Umeå University, Sweden.

E-mail: camilla.sandstrom@pol.umu.se.

Camilla Sandström is Senior Lecturer and Associate Professor at the Department of Political Science, Umeå University, Sweden. She is currently engaged in the research program Future Forests, focusing on forest policy and governance. Professor Sandström has a special interest in approaches to conflict management, and together with colleagues she has developed methods to manage forest- related conflicts. At present Dr. Sandström is also leading an interdisciplinary team involved in research on sustainable rural development with a specific emphasis on natural resource management such as protected areas, forest, fish, and wildlife.

Anna Sténs

Researcher in History, Department of Historical, Philosophical and Religious studies, Umeå University, Sweden.

E-mail: anna.stens@historia.umu.se

Anna Sténs is a researcher and teacher in history and environmental history at the Department of Historical, Philosophical and Religious Studies, Umeå University, Sweden. She is currently carrying out research within Future Forests Research Program. Her focus is on comparative studies involving Canadian and Swedish forestry and the history of forest aesthetics. Dr. Sténs has a special interest in analyzing political ideas and their implementation, and has studied national migration policies and settlement movements in the early 20th century. She is also a curator and museologist, and has worked as a consultant on issues regarding natural and cultural heritage, mapping land use history for future landscape management and documenting and informing homeowners about the preservation of historical sites and buildings.

Madelene Ostwald

Associate Professor in Geography, Centre for Climate Science and Policy Research (CSPR)/Tema V, Linköping University, Sweden, and Centre for Environment and Sustainability (GMV), Chalmers University of Technology, Göteborg, Sweden.

E-mail: madelene.ostwald@liu.se

Madelene Ostwald is an Associate Professor in Geography, especially physical geography, with a research focus on land use and forestry in relation to changes climate change and associated policies, where studying different aspects of deforestation, through the REDD+ initiative, has been central. Carbon, as the central climate commodity, has been assessed through field measurements and calculations. Furthermore, Prof. Ostwald has carried out assessments through remote sensing, analyses of accounting related to international carbon leakage and impact assessments of REDD+ pilot regions in tropical countries. She works at the Centre for Climate Science and Policy Research (CSPR)/Tema V at Linköping University and at the Centre for Environment and Sustainability (GMV) located at Physical Resource Theory, Department of Energy and Environment, Chalmers University of Technology, Göteborg.

List of Figures

List of Photos

List of Tables

Chapter 1
Nordic Forest Futures – An Introduction

Karin Beland Lindahl, Erik Westholm, and Florian Kraxner

This book focuses on how global trends are likely to affect the future use of Nordic forests. The aim is to contribute to a broad debate about future Nordic forest management. The book invites professionals in the forest sector, civil society organizations and decision makers to be part of a dialog about the opportunities, challenges, and trade-offs associated with future forest use. The book is produced within the Future Forests Research Program (www.futureforests.se), a major cross-disciplinary research effort to address future Swedish forest use in the light of climate change and an increasing demand for forest-related products and services.

In eleven chapters, well known authors explore different global trends and their possible implications for Nordic forest management. The book is basically a story about change. However, it is not the first attempt to explore change in relation to forests and forest use. In the early 1990s, Alexander Mather released his forest transition theory (Mather 1990, 1992). This focuses on how *forests* change, more specifically how a country's forest cover is likely to decline as the country develops socially and economically and how thereafter the forest cover reverses and expands. The result is a "U-shaped" curve for forest cover as a function over time. Other social scientists have followed Mather in searching for evidence of a forest transition, and the theory has been debated for decades. More recently, the term forest transition has been used to describe the transition from production to post-industrial or

K. Beland Lindahl
Unit of Political Sciences, Luleå University of Technology (LTU), Luleå, Sweden

E. Westholm (✉)
Department of Urban and Rural Development, Swedish University of Agricultural Sciences (SLU), Uppsala, Sweden
e-mail: erik.westholm@slu.se

F. Kraxner
Ecosystems Services and Management Program (ESM),
International Institute for Applied Systems Analysis (IIASA), Laxenburg, Austria

E. Westholm et al. (eds.), *The Future Use of Nordic Forests*,
DOI 10.1007/978-3-319-14218-0_1

Photo 1.1 Native forests in the fall

post-productive forests in developed countries. This latter form of transition implies a shift in forest policy and practice with a reduced "emphasis on timber production relative to the provision of environmental goods and services" (Mather 2001; Mather et al. 2006). One of the focuses here is on how *forest policy* changes (i.e., policy transition). Other research focuses on how the *forest sector* changes, for example, how its composition, processing chains, outputs, and trade shift in response to changing demand and consumption. Long-term forecasts of global production, consumption, and trade in forest products have traditionally been produced by institutions such as the United Nations Food and Agricultural Organization (UNFAO) and the United Nations Economic Commission for Europe (UNECE). In the 1980s the International Institute for Applied Systems Analysis (IIASA) developed a global model of the forest sector, and long-term scenarios were produced within the framework of its Forest Sector Project (Kallio et al. 1987). The chapters of this book discuss *forest transition, policy transition,* and *forest sector transformation* in an integrated manner.

What global changes are likely to affect the conditions for future Nordic forest use? Beland Lindahl and Westholm (2011) identify four areas that stand out as particularly important: changing energy systems, emerging international climate policies, changing governance systems, and shifting global land use systems. Demographic and economic developments underlie all of these. Our focus on these trends should be seen as an attempt to capture new developments that have the potential to cause major transformations of the Nordic forest sector. We explore each of them in greater depth and discuss their possible implications for future Nordic forest use. Changing forest products markets, technical innovation, and industrial restructuring are given less priority. Several studies suggest the Nordic forest sector is already transforming in response to many of these trends (Beland Lindahl and Westholm 2011; Lehtinen et al. 2004; Ottosson 2011). However, how forests and forest management will be affected depends on how the various actors and states react. Therefore, we also compare how Sweden and Canada, two major northern forest-producing nations, are positioning themselves to tackle expected change.

Nordic forests have been linked to international developments for centuries. Shifting markets, geopolitics, technological developments, and a fluctuating world economy are factors that have influenced the evolution of the Nordic forest sector. As early as the seventeenth century, European geopolitics and the need for timber and tar to keep the navies afloat had significant impacts on Norwegian and Swedish forests (Westoby 1989; Kardell 2003). Charcoal, nitric acid, tar, and potash became important Swedish export products in the eighteenth and nineteenth centuries (Östlund 1993; Lindqvist et al. 2009). Increased international demand for sawn timber, triggered by the industrial revolution in England, drove the expansion of industrial forestry to the far north of Sweden by the end of the nineteenth century (Kardell 2003; Bäcklund 1988; Lundgren 1984; Bladh 1997). The twentieth century brought an industrial wood regime characterized by expansive mass production across northern conifer-dominated forests (Lehtinen et al. 2004). Sawn timber, wood-based pulp, newsprint, and an increasing number of paper grades were manufactured on a commercial scale. The expanding forest industry contributed to growing welfare, providing support for the development of the Nordic welfare states. Since the 1980s the Nordic forest industry has undergone a rapid process of growth, consolidation, internationalization, and ecological modernization. In the last decades of the twentieth century three major challenges faced the global forest industry: an expansion in quick-growing fiber species in the South, increased use of recycled fibers, and growing environmental concerns (Lehtinen et al. 2004). Since then, climate and energy issues have also come on to the global and Nordic forest agendas.

The future of the Nordic forest sector is of obvious interest to the Nordic states, forest owners, forest industries, and people who spend time in Nordic forests. However, there are also several reasons why the development of Nordic forestry is of interest to an international audience. First, the Nordic countries have highly international, competitive, and export-oriented forest industries. During the twentieth century, the forest industry grew into a large source of export revenue and became a major employer in Norway, Sweden, and also Finland. In 2010 Sweden and Finland accounted for 18 % of global sawn and planed softwood exports and about the same proportion of the global export of paper and paper board (Swedish Statistical Yearbook of Forestry 2012). The Nordic forest corporations, many established more than 100 years ago, are still among the largest in the world. Some of them are now investing in forest plantations and industry facilities in Africa, Asia, and Latin America. The Nordic corporations' contribution to global exports and their worldwide operations give them a significant role in the global forest sector.

The second reason for taking interest in the development of the Nordic forest sector is the forest management concept that has become known as the Nordic forestry model. This is generally associated with efficient—and sustainable—industrial forest management rooted in a tradition of consensus politics and compromise. Finland, Norway, and Sweden are all countries at a late stage of forest transition (Mather 1992), that is, they have all been able to reverse forest decline and demonstrate how it is possible to combine industrial forestry with an expanding forest cover. Mather (2001) goes as far as to suggest that Scandinavian forestry now is influenced by "post-productivism," implying there has been a shift in emphasis from production of industrial goods to the provision of services and "greening" of forest management. In spite of this "greening," Sweden and Finland have been able to develop a competitive international forest industry. The image of the Nordic forestry model as a road map to sustainable forest management has accordingly spread around the world. Through these countries' international forest industry companies, development aid cooperation, and active participation in various international forest organizations, the ideas underpinning the Nordic forestry model have diffused into many other geographical areas (Rekola et al. 2010). Sweden, for example, has a long history of forest collaboration in Africa. At the core of many development projects is the idea of transferring the Swedish experience of developing sustainable and multiple-use forest management (see, e.g., Lundgren et al. 2011a, b). Numerous delegations from China, Vietnam, and other Asian countries have visited Sweden and Finland to learn about the Nordic example, and Finnish forest consultants are exporting Nordic forest "know-how" on a global scale. Consequently, it may be argued that the influence of the Nordic forestry model is greater outside Nordic countries than inside. Meanwhile, the "Swedish forest model" is subject to growing criticism within Sweden, among other things because it has failed to maintain biodiversity as required by the national Environmental Quality Objectives (Swedish Environmental Protection Agency [SEPA] 2012). In Finland too, the balance between production and biodiversity is being debated (Forest.fi 2012).

A third reason for the attention being paid to the Nordic forestry model is its association with the Nordic welfare model. There is no established consensus about the definition, or even the existence, of a specific Nordic forestry model. However, it has often been viewed as an integrated part of a more general Nordic welfare model characterized by a specific balance between economic, social, and political interests (Donner Amnell 2004). The Nordic welfare model is sometimes described as a socialist model recognizable by a high level of resource distribution and primary service provision by the state (Esping-Andersen 1990). More common, however, is the recognition of the Nordic welfare model as a specific form of governance based on public/private cooperation and efficient vertical integration (Westholm and Beland Lindahl 2012). A prominent feature of the Nordic welfare model is the development of small and open economies with dominant export sectors based on raw material extraction (Kosonen 1993; Andersson and Mjöset 1987). The forest sector has thus been seen as a part of a welfare project "aiming to ensure equal conditions and allow citizens to control and consciously steer their own lives" (Palme et al. 2001).

Taken together, these three factors explain the interest in Nordic forest management around the world. During the last decades, the Nordic forest sector has been increasingly affected by changes outside the Nordic countries. Demographic growth, climate change, and an expected expansion in renewable energy sources are currently the main subjects in a global discussion about future land use. In this book we try to discuss these trends and their implications for the Nordic forest sector. Many of the chapters refer to Swedish examples. This is logical, given that the book is part of a research program exploring future Swedish forest use. We argue that the similarities between the different Nordic models, viewed in an international perspective, are great enough to allow a generalized discussion about common global drivers and challenges. However, this does not imply that the different Nordic countries—and forest sectors—have similar response mechanisms. As a matter of fact, quite different strategies can be discerned. With this book, we hope to stimulate the various discussions about future forest use that are evolving in the Nordic setting.

This book is part of a research program about forests, but it can also be read as a futures study. Forests and futures are naturally related. The forest sector, with planning cycles of several decades, has a long tradition of adopting a long-term perspective. Especially in boreal areas where forests grow for 80–100 years before being harvested, investments are bound to rely on long-term estimates, forecasting, and scenarios. Yet the contributions of this book are not forecasting exercises, and prediction is not their main purpose. The book is a futures study exploring possible future developments in order to highlight trade-offs and choices. The future does not yet exist. Therefore, we have no empirical data with which to establish definite scientific conclusions about it. In this sense, our exploration of possible futures may be seen as an informed way of not knowing. We acknowledge that human agency may change the course of events; we leave space for political choice and we are careful not to determine or colonize the future with our statements about it (Andersson 2008; Sardar 1999). However, choice presumes that those who are to choose understand the options, consequences, and uncertainties it involves. The objective of this book is therefore to contribute to a broad social debate about the choices and trade-offs that may be waiting.

In the following chapters, authors from various disciplinary backgrounds use a range of different approaches to explore global trends and their implications for future Nordic forest use. Different approaches are used in the chapters: stable trends (Chap. 3), scenario techniques (Chaps. 3, 4 and 5), interpretive institutional/frame analysis (Chaps. 6 and 8), policy interpretation (Chap. 7), and historical explorations (Chaps. 9 and 10).

In Chap. 2 **Erik Westholm** introduces futures studies and the challenges of this research field. He reflects on how the notion of the future has changed over past decades, the importance of adequately addressing the temporal dimension, and how we can deal with the realization that our knowledge about the future will always be incomplete and uncertain.

In the third chapter **Bo Malmberg** explores global demographic trends. Populations change slowly, and largely in ways that are easy to predict. Malmberg demonstrates how population projections can be used to forecast future economic growth and demand for natural resources in various social settings. He uses the term demographic transition to describe the long-term process by which a country's birth and death rates fall from high to low levels, as it develops economically. The timing of this transition in different parts of the world is key to understanding how the demand for forest products will shift in the future. Malmberg shows that the demand for forest products is likely to increase dramatically in the coming decades. In line with Mather (1992), he discusses two interrelated trends: demand driven depletion versus forest cover recovery due to increased productivity and improved management practices. According to Malmberg, demographic changes are likely to drive a large-scale transition from natural forests to intensive forest plantations.

Demographic growth will boost demand, not only of forest products but of a range of commodities that require land for their production. Numerous studies suggest that expected demographic and economic growth, under the pressure of climate change and possible energy supply shortages, may lead to increased global land use competition. In Chap. 4 **Sten Nilsson** reviews state-of-the-art scenario analyses developed to explore complex and integrated land use systems. He discusses the integrated and accumulated global land use impacts of an increased demand for energy, food, and forest products under the climate scenarios developed by the Intergovernmental Panel on Climate Change (IPCC 2007). Nilsson highlights the risk of global land shortages and stresses the need for timely and efficient governance interventions. He concludes by discussing the implications for the Nordic forest sector of intensified global land use competition.

One key variable in the global land use equation is the development of bioenergy feedstock production. This, in turn, is driven by a global demand for renewable energy which is expected to increase as a result of climate mitigation strategies and a diminishing supply of conventional fossil fuels. In Chap. 5 **Florian Kraxner** and **Eva-Maria Nordström** explore different global scenarios for production of bioenergy feedstock under specified social and environmental safeguard provisions. Results indicate that with rising populations and projected consumption levels, there may not be enough land simultaneously to completely conserve natural forests, halt forest loss, and switch to 100 % renewable energy, which will make difficult trade-offs necessary. Kraxner and Nordström suggest that managed boreal forests are likely to become an important source of bioenergy feedstock.

So far, the different chapters have painted a broad picture of a number of interrelated trends and challenges that are emerging at the global level. However, they all underline that the difference between disaster and sustainability in 2050 lies in the speed and efficiency of implementing the right policies. In Chap. 6 **Katarina Eckerberg** analyzes current global trends for forest governance. More specifically, she explores the interaction between the forest sector and other related policies/sectors and their influence on social and political institutions for forest management. Her main argument is that the institutional governance framework is changing rapidly and that the global changes taking place will affect the forest governance situation in the Nordic countries. New conflicts between interests and accompanying novel alliances are likely to emerge. A burning issue is the role of the nation state; policy instruments and governance structures are seemingly moving away from the national level but are yet to be replaced by alternative transparent and democratically legitimate institutions.

Climate-related forest policy is a growing policy field witnessing a number of emerging global governance mechanisms. In Chap. 7 **Madelene Ostwald** explores the evolution of global climate-related forest policies, such as the Clean Development Mechanism (CDM), the REDD+ process, the voluntary carbon market, and climate-related management strategies for multiple benefits. Ostwald describes the role of forests in various carbon accounting and carbon mitigation schemes and discusses their challenges and opportunities. Forest-related carbon is increasingly becoming a tradable commodity along with other products and services that the forests supply. Although the main focus of the international climate negotiations has been on tropical forests because of their high contribution to global greenhouse gas (GHG) emissions through deforestation, the emerging international agreements will have implications for the Nordic forest sector. Voluntary markets for forest-related carbon offsets will, for example, present Nordic forest owners and carbon emitters with new options and incentives.

In Chap. 8 **Karin Beland Lindahl** explores global change from a more specific Swedish perspective. She uses Sweden as a point of departure for an analysis of how key actors perceive the challenges and opportunities of global climate-, energy-, and demography-related trends. Actors' perceptions of future challenges and opportunities influence their choice of strategy and action. Actors' relative capacity to achieve their visions, in turn, shapes future forest use. Whereas Chap. 6 focuses on institutional change, Chap. 8 explores how actors' responses to global change affect forest governance. Beland Lindahl uses frame analysis to identify major divisions and competing pathways to "sustainability." A major division separates actors who perceive biomass supply as unlimited, or at least as not constraining, and those who stress scarcity and redistribution of resources. This difference, or frame conflict, is reflected in actors' forest-related strategies and may fuel future forest conflicts in the Nordic region.

In Chaps. 9 and 10 we look more deeply into possible responses to the global challenges outlined in previous chapters. How the global trends will materialize in the different Nordic countries depends on institutional innovation, market developments, and policy decisions reflecting the preferences, strategies, and relative strengths of the actors in the political arena. How well the sector manages to respond, that is, to transform in order to meet the global challenges in a proactive way, will determine its future position in the international market and forest management discourse. All northern forest-producing nations face similar global pressures, and a transformation of the sector is already under way in many places. In Chap. 9 **Sten Nilsson** offers an "insider's" account of how the Canadian forest sector is transforming to meet structural challenges and global change. Canada is particularly interesting, as serious structural problems and environmental conflicts have already triggered a broad and concerted initiative to renew the Canadian forest sector. Canada is a boreal forest nation which shares many similarities with the Nordic countries, although differences in land ownership, forest tenure, etc., have to be taken into account in any comparison made. Nilsson discusses the problems that have triggered the Canadian transformation process and introduces two vital components: the Canadian Boreal Forest Agreement and the Future Bio-pathways Project. He concludes with a discussion of commonalities and differences between the Canadian and Swedish forest sectors. Whereas some features of the Canadian process are context-specific, other experiences are clearly relevant to the Swedish and Nordic forest sectors.

In Chap. 10, **Camilla Sandström** and **Anna Sténs**, explore "the Swedish forest model" and how the Swedish forest sector has responded to previous pressures to transform. They analyze past dilemmas encountered in the governance of policy transitions with a view to exploring the potential for managing a long-term socio-ecological transformation of the Swedish forest sector. Sandström and Sténs pose two critical questions: to what extent is it possible to govern forest policy transitions? and, is it possible to influence the direction and speed of a transition while resolving trade-offs arising from the transition process? Based on their analysis, the authors conclude that although the Swedish government has tried to deal with several of the dilemmas identified, it has not managed to handle the most fundamental one, that is, bridging the gap between key stakeholders through continuous and iterative deliberation. In contrast to the Canadian example described in Chap. 9, the gap between Swedish key actors currently seems to be too wide to expect any joint effort to transform the sector.

The book moves from an international perspective on demography-, climate-, and energy-related trends to an exploration of possible Nordic responses to these challenges. In the concluding discussion, we elaborate on the main challenges facing the "sector": the choices, trade-offs and governance challenges that the actors must be prepared to deal with. The future of the Nordic Forest Model is of high international relevance.

References

Andersson, J. (2008). *The future landscape* (Arbetsrapport 2008:5). Stockholm: Institute for Future Studies.

Andersson, J. O., & Mjöset, L. (1987). The transformation of the Nordic models. *Cooperation and Conflict, 22*(2), 227–243.

Bäcklund, D. (1988). *I industrisamhällets utkant: Småbrukets omvandling i Lappmarken 1870–1970* (Umeå studies in economic history 8). Doctoral thesis, Umeå University. [In Swedish]

Beland Lindahl, K., & Westholm, E. (2011). Food, paper, wood or energy? Global trends and future Swedish forest use. *Forests, 2*, 51–65.

Bladh, G. (1997). Gruvved och kolskog. Skogens utnyttjande i Bergslagen perioden 1500–1900. In L. Östlund (Ed.), *Människan och skogen: från naturskog till kulturskog?* (Skrifter om skogs- och lantbrukshistoria 1, pp. 85–103). Stockholm: Nordiska museet. [In Swedish]

Donner-Amnell, J. (2004). To be or not to be Nordic? How internationalization has affected the character of the forest industry and forest utilization in the Nordic countries. In A. Lehtinen, J. Donner-Amnell, & B. Saether (Eds.), *Politics of forests: Northern forest – Industrial regimes in the age of globalization* (pp. 179–204). Aldershot: Ashgate.

Esping-Andersen, G. (1990). *The three worlds of welfare capitalism*. Cambridge/Princeton: Polity Press/Princeton University Press.

Forest.fi. (2012). *Revised Forest Act to diversify forest management practices*. http://www.forest.fi/smyforest/foresteng.nsf/allbyid/AF073CA03690ACCBC2257A60003CEC6C?OpenDocument. Accessed 22 Nov 2012.

IPCC. (2007). Climate change 2007: Synthesis report. In R. K. Pachauri & A. Reisinger (Eds.), *Climate change and its impacts in the near and long term under different scenarios* (pp. 21–54). Geneva: IPCC.

Kallio, M., Dykstra, D. P., & Binkley, C. S. (Eds.). (1987). *The global forest sector: An analytical perspective*. Chichester: Wiley.

Kardell, L. (2003). *Svenskarna och skogen: D. 1, Från ved till linjeskepp* (pp. 217–223). Jönköping: Skogsstyrelsens förlag. [In Swedish]

Kosonen, P. (1993). The Finnish model and the welfare state in crisis: The Nordic welfare state as an idea and as reality. *Renvall Institute Publications*, 5, 45/66.

Lehtinen, A., Donner-Amnell, J., & Saether, B. (2004). Introduction: Northern forest regimes and the challenge of internationalisation. In A. Lehtinen, J. Donner-Amnell, & B. Saether (Eds.), *Politics of forests: Northern forest-industrial regimes in the age of globalization* (pp. 3–30). Aldershot: Ashgate.

Lindqvist, A., Mineur, E., Nordlund, A., Nordlund, C., Olsson, O., Sandström, C., & Westin, K. (2009). *Konflikt och konsensus: Intensivodling av skog ur ett humanistiskt och samhällsvetenskapligt perspektiv; faktaunderlag till MINT-utredningen*, SLU Rapport (pp. 8–9). Umeå: Swedish University of Agricultural Sciences. [In Swedish]

Lundgren, N. G. (1984). *Skog för export: skogsarbete, teknik och försörjning i Lule älvdal 1870–1970* (Umeå studies in economic history 6). Doctoral thesis, Umeå University, Umeå, Sweden, 1984. [In Swedish]
Lundgren, B., Persson, R., & Norén, S. (2011a). Swedish-African forest relations. *Kungl. Skogs- och Lantbruksakademiens Tidskrift,* nr. 2. Stockholm: Royal Swedish Academy of Agriculture and Forestry.
Lundgren, B., Owino, F., Ishengoma, R., & Gondo, P. (2011b). *African-Swedish collaboration programme on sustainable forest management*: Final report on the planning phase of the project. Nairobi, Kenya: The African Forest Forum.
Mather, A. (1990). *Global forest resources*. London: Belhaven Press.
Mather, A. (1992). The forest transition. *Area, 24*, 367–379.
Mather, A. (2001). Forests of consumption: Postproductivism, postmaterialism, and the postindustrial forest. *Environment and Planning C: Government and Policy, 19*(2), 249–268.
Mather, A., Hill, G., & Nijnik, M. (2006). Post-productivism and rural land use: Cul de sac or challenge for theorization? *Journal of Rural Studies, 22*, 441–455.
Östlund, L. (1993). *Exploitation and structural changes in the north Swedish boreal forest, 1800–1992*. (Dissertations in forest vegetation ecology 4, pp. 13–16). Umeå: Swedish University of Agricultural Sciences.
Ottosson, M. (2011). *Opposition and adjustment to industrial "greening": The Swedish forest industry's (re)actions regarding energy transition: 1989–2009*. Doctoral thesis, The Department of Thematic studies: Technology and Social Change, Linköping University, Sweden.
Palme, J., Bergmark, Å., Bäckman, O., Estrada, F., Fritzell, J., Lundberg, O., Sjöberg, O., Sommestad, L., & Szebehely, M. (2001). *Welfare in Sweden: The balance sheet for the 1990s*. Stockholm: Fritzes.
Rekola, M., Valkeapää, A., & Rantala, T. (2010). Nordic forest professionals' values. *Silva Fennica, 44*(5), 885–908.
Sardar, Z. (1999). The problem of future studies. In Z. Sardar (Ed.), *Rescuing all our futures: The future of futures studies* (pp. 9–19). Twickenham: Adamantine Press.
Swedish Environmental Protection Agency (SEPA). (2012). *Swedish statistical yearbook of forestry*. 2012. Internationell Skogsstatistik. http://www.miljomal.nu/. Accessed 27 Nov 2012.
Westholm, E., & Beland Lindahl, K. (2012). The Nordic welfare model providing energy transition? A political geography approach to the EU RES Directive. *Energy Policy,* 50, 328gy P.
Westoby, J. (1989). *Introduction to world forestry: People and their trees*. Oxford: Blackwell.

Chapter 2
Futures Studies in the Field of Natural Resources

Erik Westholm

This chapter reflects on futures studies and the role that "the future" plays in social and economic transformations. The aim of the chapter is to go into some detail on the history of futures studies, with emphasis on the futures envisaged in the forest sector and in the field of the environment since the 1950s. It situates the role of the "future" in today's environmental debate referring to dialogs that have taken place between productivist and environmental discourses.

2.1 Introduction

The authors of this book were asked to take a long-term perspective when exploring the future of Nordic forests. The forest sector, which has planning cycles spanning several decades, has a long tradition of engaging with the distant future. Forest management in the Nordic countries has been concerned with the future consequences of present practices for the last 200 years. Indeed, investments, especially in boreal areas, where forests grow for 80–120 years before being harvested, are bound to rely on long-term estimates, forecasting, and scenarios.

In the book we link these characteristics of Nordic forestry to another set of long-term issues: the global challenges, foreseen during the last few decades, that result from demographic and economic growth, climate change, and the crucial transition

E. Westholm (✉)
Department of Urban and Rural Development, Swedish University of Agricultural Sciences (SLU), Uppsala, Sweden
e-mail: erik.westholm@slu.se

E. Westholm et al. (eds.), *The Future Use of Nordic Forests*,
DOI 10.1007/978-3-319-14218-0_2

Photo 2.1 Futures studies – a scientific approach to imagination

to a post-fossil energy system. The book is a futures study in that it aims to raise awareness in the forest sector of both the constraints and the opportunities facing Nordic forests.

From a methodological point of view the chapters of the book vary significantly, reflecting the fact that futures studies is a scattered scientific field. In their chapter on "Bioenergy futures," Kraxner and Nordström provide a classic modeling approach that is able to explore different possible outcomes when data and external conditions are varied in the model. Malmberg makes forecasts based on demographic change by utilizing *a theory of social and economic change*. Nilsson's analysis of land use futures is based on a synthesis of many sources, *a compilation of futures studies*. This is very different from his participatory approach in the chapter on the Canadian forest transformation, which is built on an *actual transformation of the Canadian forest sector*. Beland Lindahl explores the future in an *interview-based study on perspectives and strategies among key actors,* while Eckerberg applies a *political science approach* when discussing governance opportunities in the future. Ostwald's contribution addresses the outlooks resulting from the new steering mechanisms in *forest-related climate politics*. Finally Sandström and Sténs make their assessment of *the future of the Swedish Forestry model* in the light of global challenges.

These contributions have at least one methodological attribute in common; they are not trying to "discover" a pre-existing future. The purpose is rather the opposite: to start an informed discussion in the present on potential solutions to the grand challenges related to energy, climate change, and land use. Possible future developments are explored with the aim of highlighting the choices to be made and the trade-offs. "The future" is seen as relatively open to human agency.

As the long-term perspective is one of the book's main pillars, the book must be clearly positioned in the field of futures studies. A meta-discussion on "the future" is important because of the exclusive nature of futures studies as a scientific field. As the future is nonexistent, it must be constructed in the present base on conclusions from transformations of the past. To reveal "the hidden pulse of history" is the core of futures studies. Yet, historical analysis alone is not sufficient. The future is continually constructed as the result of insights, perspectives, and strategic interests among more or less influential actors. Therefore, the notion of the future, and of the future as constructed in the present, needs to be communicated among professionals dealing with the future in research and in politics.

The contribution of this chapter is to reflect on futures studies and the role that "the future" plays in social and economic transformations. For many decades, the future has been an organizing concept for debates on development and modernity; notions on the future have reflected visions and dystopias that vary over time (Anderson 2011).

Three phases of futures studies are identified;

(i) After World War II the first systematic futures studies emerged, closely linked to the progress of industrialism. Optimistic futurism, or futurology, developed as a scientific field. The core of these futures studies was to improve the tools for prediction.

(ii) The arrival of environmental debates during the 1960s and 1970s introduced a democratic approach to futures studies. It was based on the view that the future is yet to be decided, and can and must be consciously chosen in democratic processes. The core of these futures studies was a view of the future as being open to human agency.

(iii) From the 1990s the idea that the future can be deliberately steered for the better was challenged by ideas of risks and uncertainty. This shift produced a new futures horizon that was increasingly concerned, not with planning for one or several futures, but preparing for unexpected developments. The core of these futures studies has been the notion of risk and uncertainty.

In the end it must be recognized that the different notions of the future have not replaced each other but developed gradually as complements forming an increasingly complex relation to the future.

2.2 From Embedded Time to Symbolic Time

The future is an essential part of any human activity, and all societies develop their own ways of coming to terms with it. These knowledge processes have multiple functions; they help structure social life and make it possible to anticipate regular and irregular patterns in society and in the nature. In pre-industrial societies, learning about the future was mainly based on understanding the repeated cycles in nature. Knowledge of the movements of the planets, climatic circles, seasonal variations, etc., enhanced predictability beyond the immediate present (Adam and Groves 2007). The expected continuity was transformed into social regularities; rituals and festivals helped anchor social life in patterns of anticipated events. Adam and Groves have made two important statements in relation to these pre-industrial futures: (i) they were embedded in place- and time-specific conditions, that is, they were localized and culturally rooted; and (ii) knowledge of them could be transferred between generations in the local society.

In the early phases of European industrialization in the eighteenth and nineteenth centuries, it was already centuries since consciousness of a world beyond the immediate space of everyday activities had started to transform perceptions and strategies in relation to the future. Money brought symbolic values into trade relationships and the concepts of interest and credit caused a definite shift in social, spatial, and temporal relationships. With increased long-distance trade, social/economic relationships were lifted and gradually disembedded from the local. As a result, notions of the temporal dimension changed. As Adam and Groves (2007, p. 10) put it:

> At issue is no longer the contextual, embodied future, but a future emptied of content and divorced from context, a future that can be calculated anywhere at any time and exploited for any circumstance.

What Adam and Groves described here is a future that becomes visible as a separate dimension, something detached from ourselves and our everyday lives. What the authors describe as a "future emptied of content and divorced from context" brings to mind a discussion by Swedish geographer Torsten Hägerstrand (1985) in which he usefully distinguishes between "embedded time" and "symbolic time." Embedded time is lived time, the daily practice in which time is intimately interwoven, taken for granted, and invisible, and therefore cannot be named or reflected upon. Symbolic time, in contrast, is what we call time and what enables discussion of the past, present, and future (Wästfelt 2006).

Symbolic time is the conceptual tool we use when observing changes. However, the distinction between embedded time and symbolic time is not clear-cut. The further we move away from activities rooted in everyday life. the more relevant the notion of symbolic time becomes. In forestry, with turnaround cycles ranging over generations, such a meta-reflection on time was already at hand during the proto-industrial era of the eighteenth century and became central to forest debates during the industrial expansion. The concept of *Nachhaltigkeit*—sustainability—had already been coined in the mid-nineteenth century to describe the eternal cycle of forest production, in which needs were considered generations into the future (Hölzl 2010). Obviously, too, the future that was envisaged was not an embodied future based on continuity and repetition but on expectations of further growth. This was more than a century before the sustainability concept was introduced into the environmental debate with its now ecological connotations.

2.3 Futures Studies in the Post-War Era

The post-World War II period is usually described as the time when scientific futures studies emerged. After the war, and with the Soviet Union and United States (USA) as pioneers, rapid technological development increased the interest in long-term planning, forecasting, scenario techniques, etc. (Skovdahl 2012). These attempts to forecast the future were largely instrumental and never developed into an epistemological basis for scientific approaches. In the 1950s they were developed as forecasting techniques, often concerned with trend projections in relation to a growing economy and expanding production of services and goods. The idea that the future could be scientifically explored by projections generated a perception of the future itself as the product of technological progress (Andersson and Keizer 2012). The futures debate of the 1950s was dominated by the idea of the continued progress of welfare in Western economies. At times, the visions were disrupted by dystopian ideas related to fears of devastating warfare resulting from the development of nuclear technologies.

In the 1960s and 1970s, the RAND Corporation and the Hudson Institute in the USA dominated the development of a "futurology," of which Herman Kahn and Daniel Bell were leading exponents. *The coming of the post-industrial society* (Bell 1999), which was influential for several decades, was based on the notion of scientific progress and its industrial applications as the drivers of modern society. The main interest in Bell's work, however, was not its analysis of technological progress as such, but its reading of the social consequences of this progress.

These works were directed toward improving predictions of the future, which was presented as something phenomenological, to be captured and described by futurologists. Modeling techniques were improved with quantitative methodologies and massive amounts of data. The work of Kahn and of Bell expressed the optimistic futurism that dominated the period. It presented the various stages of human development and progress, from the early phases of agricultural production, through the arrival of industrialism, to the contours of a post-industrial society.

After World War II, the ideas of the general Nordic welfare model could be fully developed. The Nordic welfare model was a modernization project linked to the shift of the labor force from the primary sector to the growing industries. The ambition was to provide equal conditions for all households in all parts for the country, while at the same time safeguarding equal conditions for individuals and households (Westholm 2010). Increased household purchasing power should function as a driver of economic growth and exports.

The forest sector, with its rapidly expanding pulp and timber industries, was an important producer of both the employment opportunities and export revenues needed for the welfare project. In Sweden, a new forest policy was directed toward increased production to meet future product demands and to provide economically for the growing welfare. Through the Swedish Forest Conservation Act of 1948, forestry became an integrated part of the modernization project. One aim was to strengthen the institutional base through a reinforcement of the National Forestry Board and the formation of a wide network of authorities and experts to implement the new policy. These attempts were based on a consensus within the forest community on both the theoretical goals and the means to achieve them: a trust in increased production following from successful R&D. In state politics, the efforts to improve the forest sector was linked to the establishment of the Nordic welfare model with its ambition, as mentioned above, of pushing industrial modernization while at the same time safeguarding equal conditions for individuals and households. The establishment of the Nordic welfare model and post-war forest politics is an example of what Scott (1998) calls "high modernism"— basically, large-scale efforts to use science and technology to order and standardize the social and natural world, and to use notions of the future to reshape the present.

2.4 Environmentalism and the Rise of "Democratic Futures Studies"

Another fertile strand of futures studies emerged in the late 1960s with the swelling popular movements concerned with peace and environmental politics. Rachel Carson's (1962) book *Silent spring* with its dystopian picture of the future, triggered a heated debate about the use of chemicals in forestry and agriculture (Lisberg

Jensen 2006). A few major scenario studies gave rise to environmental nongovernmental organizations (E-NGOs) which advanced a view of the world as a common global and vulnerable system (Anderson 2012). Not least, the Club of Rome report (Meadows et al. 1972), *The limits to growth*, triggered a worried discussion over future demographics and economic growth. As well as its attempt to forecast the future, the report also generated debate on the role of futures studies in society. This debate was highly instructive regarding the creation of futures studies in Sweden and other countries in the early 1970s, and initiated the structuring of a new area of strategic environmental research. Questions of the long term were integrated into this research landscape from the outset.

Robert Junk and Johan Galtung were leading figures in promoting the view that futures studies must be integral to democratic processes. They challenged the dominating idea of futures studies as a science of prediction and replaced it with the idea that futures studies should provide alternatives and facilitate future governance. Anderson (2012) has described the range of attempts made to establish futures studies as a scientific discipline at conferences and by establishing journals such as *Futures*. Common to these "democratic" futures studies was their connection to a rising concern for the environment. The environmental threats called for alternative modes of production and consumption.

A number of research institutions based on these new futures studies were also established in Europe in the 1970s. One of those, the International Institute for Applied Systems Analysis (IIASA), established in 1972, has since played a world-leading role in forest-related futures studies. The Institute for Futures Studies (IFS) in Stockholm was initially part of the Swedish government's cabinet office and was established to develop democratically based attempts to address the future (Lundqvist and Carlsson 2004). The IFS was involved at an early stage of forest-based futures studies (Wirén 1985).

Some 150 years after the first use of the concept *sustainability* in the forest sector, the term was reintroduced during the 1980s in environmental politics, but with a different and much broader interpretation. It followed on from the classic Brundtland report which refers to "meeting the needs of today while protecting the rights of future generations" (*Our common future* 1987; Niestroy 2005). The Brundtland version of the sustainability concept recognizes ultimate biophysical limits to growth and acknowledges the responsibility of present generations to future generations. A Brundtland account of sustainable development challenges the traditional growth paradigm and addresses questions of justice and distribution of wealth between North and South (Baker 2006).

The environmental dialogs triggered by *Silent spring* and, later, The Club of Rome and Brundtland reports, transformed the conception of "the future" as linked to development and modernity. They opened a discussion on various pathways to the future; natural resources were frequently discussed in the form of scenarios, expressing that making choices was both necessary and possible. Productivist approaches were complemented and challenged by a number of environmentalist approaches to natural resource management. In Sweden, the nuclear power issue mobilized all political parties, E-NGOs, think-tanks, etc., in debates, cultural production, publications, adverts; and the referendum held in 1978 on the future of the energy system turned into a nationwide argument on what kind of modernity to head for.

Environmentalism also came to challenge the highly productive Nordic forest sector. Its emphasis on even-aged monocultures with intensive management methods was increasingly questioned, being described as unnatural and sometimes as a road to ecological disaster. Politically, the position of the forest sector had been weakening since mechanization, and specialization had gradually reduced the labor force both in the forests and in the pulp and timber industries, thereby decreasing the political significance of forest-related activities during the 1970s and 1980s.

The Wirén (1985) study was an early imprint of this critical approach, leading to Sweden's 1993 Forestry Act. The Act stated that the forest as a national asset should be managed in a sustainable way ensuring good productivity while maintaining biodiversity (Regeringens Proposition 1992/93: 226, p. 3). The Act was based on the assessment that recent decades had been successful in increasing standing forest volumes and that, at the same time, the provision of forest products had added to the growth of the forest industry. Now forest management practices should also take environmental consideration and include biodiversity and esthetic and cultural values. This was an important step in the development of the Nordic forestry model, which has been recognized mainly as a compromise between productivist and environmental objectives.

However, the Nordic welfare model with its significantly high degree of state intervention was already in retreat (Palme et al. 2001) when the Nordic forestry model came through politically. Thus, while the 1993 Forestry Act endowed forestry with more complex political objectives, it was accompanied by extended private property rights; landowners could now more freely interpret and transform the new forest policy into forest management practices. The new orientation of forest policy rested on expectations that landowners are the appropriate caretakers of both productivist aims and environmental concerns. The image of the Nordic forest model as a road map to sustainable forest management spread around the world.

2.5 The Twenty-First Century: Featuring Uncertainty and Risk

The last few decades, have seen another fundamental shift in the general perception of "the future." Previous attempts to create an anticipatory basis for action have been transformed by a general demise of ideas on public planning and state governing processes. Already at the beginning of the 1990s, Beck (1992) and Giddens (1990, 1994) had developed the idea of a "risk society," based on a discussion of a new form of modernity. Beck defines a risk society as a systematic way of dealing with hazards and insecurities induced and introduced by modernization itself. The *risk* concept aims to capture the transition from the industrial society, with its characteristic ideas on mastering nature, to a reflexive modernity, increasingly occupied with handling the negative consequences of the industrial society. Giddens (1999) identifies a society increasingly preoccupied with the future, and with safety, which gives rise to the notion of risk. As pointed out by Lakoff (2007) and Giddens (2009), the risk society thesis leads to a different future horizon or future rationality, one that is no longer about planning for security and welfare, but preparing for uncertainty, risk, or even devastating future developments.

Over the last few decades, methodological developments in both the natural and social sciences have changed the notion of the future and how it should be addressed. Modeling techniques have developed rapidly and the global challenges now recognized are themselves transforming the basis of futures studies. The climate scenarios provided by the Intergovernmental Panel on Climate Change (IPCC) are a prominent example. Moreover, research institutions have been established across traditional scientific boundaries and ideas are being developed based on new ways of approaching interactions between humans and nature. Briefly stated, the agenda seems to be moving from one of establishing control and certainty to one of addressing uncertainty.

Social sciences have generally anticipated nature as a rather stable point of departure: natural processes have often been seen as linear and predictable. In these new research agendas dominated by interactions between social and natural systems, nature stands out as reactive and unpredictable. Flooding, storms, droughts, and diminishing biodiversity, as well as worrying data on global warming, are making their imprints on research agendas by favoring problem-based knowledge production. Emerging research institutions are being established based on recognizing increasingly complex human–natural interactions, and a new terminology indicates the development of ascendant interdisciplinary research fields. Social systems transition theory (Scoones et al. 2007), system flips (SEAC 2007), planetary boundaries (Rockström et al. 2009), ecological footprints (Ewing et al. 2010), soft collapse (Hornborg 2010), and third natures (Sassken and Dotan 2011) are all concepts that aim to capture the present dynamic character of environmental studies. Uncertainty and unexpected transitions are common themes. The futures produced within this broad field of "risk perspectives" are typically marked by long-term perspectives in the analytical framework, while the policy implications are marked by immediate urgency.

Climate change poses a fundamental time-related challenge, as costly actions must be taken in the present to prevent problems in the future—beyond the time horizon that human agency normally grasps. Fundamental to the challenges of climate change is that the advantages of using fossil fuels and releasing greenhouse gases are accrued in the present, while the costs are externalized, located in a distant future. The expected consequences are not recognizable and are "unexperienced"; they exist as an emerging insight that comes from scientific knowledge production rather than lived experience. Adding to that, the social and economic consequences of climate change are fundamentally uncertain, and there are few historical references to facilitate the outlooks.

The Swedish research program, Future Forests 2009, funded jointly by the forest industry, the strategic foundation Mistra, the Swedish University for Agricultural Sciences, and Umeå University, provides an interesting, and perhaps typical for the period, view on the utilization of natural resources. It clearly expresses the hopes for increased production that have dominated the forest sector for 150 years. Yet it also addresses the risk agenda of the twenty-first century: the trade-offs that will have to be made in relation to biodiversity; the uncertainties related to expected climate change; and the expectations for new business opportunities related to climate politics, energy transitions, and growing global demands for forest products. The program plan expresses a wish to take on board the grand challenges and the uncertainties and to transform them into the basis for another round of productivism.

One of the studies in the program addresses the different perspectives and strategies among actors in the forest sector in relation to the future. Researchers, forest company directors, public servants, or NGO actors, could be grouped into two main discursive pathways that seemed to produce fundamentally different ideas on current politics, the possible future development of the problems, and what should be done in relation to them (Beland Lindahl and Westholm 2011). One of two main schools of thought holds that there is an *abundance* of natural resources globally and that environmental problems and resource shortages can be solved by technological solutions and measures to enhance efficiency in consumption and production. Related to this perspective is the view that climate change is either not a pressing problem or that it is a problem that can be solved by technological progress. Other perspectives are more related to expected long-term *scarcity*, which can only be resolved by a more radical transformation of production and consumption systems. The scarcity may result from an actual lack of natural resources or from expectations that these resources cannot be efficiently mobilized. Related to this view are worries that current policies to address climate change are severely deficient.

2.6 Conclusions

Our notion of the future has arguably been transformed over time from one that is deeply embedded, place-specific, and rooted in everyday relationships to something that is more abstract and symbolic, although still important in steering human agency. In this chapter the changing notion of the future has been divided into three main periods after World War II.

During the first decades after the war, the first systematic futures studies developed, intimately linked to the progress of industrialism. Optimistic futurism, or futurology, emerged as a scientific field aiming to improve the tools for prediction. Methodologies were focused on modeling techniques and massive amounts of quantitative data. In forestry, the state had a prominent role in supporting this long-term view and strove, through regulations and monitoring, to extend the time horizons in the sector to encompass the turnaround cycles of the boreal forests. This required an insight in the need for investments in the reproduction of forests in harvested areas, despite the fact that the next harvest could be expected to be 80–100 years into a distant future.

The arrival of environmental debates during the 1960s and 1970s had a fundamental impact on perceptions of the future. The basic idea that the future is something that exists "out there" and can be predicted was challenged by the view that the future is yet to be decided, and that it can and must be consciously chosen in democratic processes. Conflicting interests in relation to the future were recognized and the growth paradigm was challenged by environmental approaches to natural resource management. During the 1980s the concept of sustainability was reinvented, now with reference to ecological, social, and economic dimensions, emphasizing the role of futures studies to provide support to long-term politics and planning. From a methodological point of view this is reflected in the frequent use of scenario-based futures studies and an emphasis on communication. State intervention declined and the trade-offs between productivist and environmental objectives in forestry were increasingly placed on the landowners.

From the 1990s another shift in the perception of the future has taken place. The idea that the future can be deliberately steered for the better, which has been an essential part of modernity, was challenged by ideas of risks and uncertainty. This shift has produced a new futures horizon which is increasingly concerned, not with planning for one or several futures, but with preparing for unexpected developments. The grand challenges are presenting themselves as threats beyond the time horizons normally built into decisions in politics, planning, investment circuits, etc. The response in the forest sector has been to produce cross-disciplinary research aimed at bringing the risk-based agenda, primarily with respect to climate change, in keeping with the seemingly timeless ambition to intensify production.

The changing perceptions and strategies in relation to the future have been presented as specific for certain periods, as if mutual substitution were, in fact, possible. In reality, however, these changing views have emerged as mutually complementary, forming an increasingly complex relation to the future.

Constructions of the future are always based on attempts to interpret and organize patterns from the past to reveal "the hidden pulse of history." These transformations of historical developments into possible futures are carried out by actors who are themselves carrying values and interests in relation to the future. In fact, dialogs over the future are always part of a struggle over ideas and values in contemporary society. Those who have the power to define the future, or colonize it with their ideas, become influential in their own time (Sardar 2010).

Over the last 50 years or so, lasting debates have been organized around the fundamental requisite of economic growth with respect to issues of scarcity versus abundance of natural resources and the externalization of environmental costs by the market, etc. This reflects a growing tension around the core of modernity which contests the idea that "we" can consciously steer the world toward ever-increasing prosperity.

The Western economies have been struggling to sustain economic growth and competitiveness while at the same time addressing environmental challenges.

Environmentally sustainable practices are incorporated into politics as long as they can be resolved without redirecting the main course of the liberal economy. Challenges and needed transitions that may be recognized as necessary, but threaten the current social and economic order, tend to be externalized in time, for example, "until 2050" or "beyond the purview of today's decisions." For instance, the politics of climate change has been divided into, first, strategies that are obviously insufficient to address the problem as it is described and recognized, and, second, vague acknowledgments of the need for much more radical political change to be addressed in a distant future. Thus, the future, or the temporality, is used as a tool to put political issues on and off the agenda, respectively.

The concept of *sustainability* expresses the hope for a political solution to the dilemma of economic growth and resource depletion. In politics, the idea of sustainable development has been transformed into a possible compromise: environmental sustainability has been incorporated into the growth paradigm and is now central to the idea of continuous growth. Moreover, the notion of risk and uncertainty tends to be incorporated into the economy. Pellizoni (2011), for example, argues that contemporary liberal governance operates through, rather than despite, disorder: that is, through contingency, uncertainty, and instability. Hence, it is speculative rather than predictive, and risk and uncertainty have been turned into an "empowering everyday condition."

When advancing a 200-year perspective, Adam and Groves (2007) highlight something central to any understanding of the changing notion of the future, namely, that industrialization was accompanied by the promise of ever-increasing incomes and wealth, which it largely succeeded in delivering during the twentieth century. The future envisaged was not an embodied future based on continuity and repeated cycles but a future filled with expected developments for the better. It has taken until recent decades for the negative consequences of those efforts to advance into a discourse dominated by threats. The grand challenges are the consequences of previous actions. Those consequences are not yet here but they are already forcing us to engage with more distant futures horizons: the *actions* (the establishment of industrialism) have become separated in time from the *knowledge* of the long-term consequences, which has in turn been separated from *the consequences* as such. Governance systems are having obvious difficulties in accounting for consequences that are not yet here.

To conclude, we can see an extension of our perceived responsibilities into a more distant future while at the same time there is an increased urgency to take actions today. The distance described by Adam and Groves (2007), the knowledge, and the consequences may already be collapsing into an immediate call for action that requires radically different modes of social organization. One obvious requisite is the need for new forms of global governance: policies able to govern a global economy. These needs and transformations are reflected in the chapters that follow on the global trends of key importance to Nordic forest use.

References

Adam, B., & Groves, C. (2007). *Future matters: Action, knowledge, ethics*. Leiden: Brills.

Anderson, J. (2011). *The great future struggle: Futures studies and the struggle for the world*. Conference paper: *Governing the world*. Sciences Po 2011

Anderson, J. (2012). En vetenskap för framtiden. Om futurologins uppkomst. In S. Alm, J. Palme, & E. Westholm (Eds.), *Att studera framtiden*. Stockholm: Institutet för Framtidsstudier och Dialogos. [In Swedish]

Anderson, J., & Keizer, A.-G. (2012, March 22–24). *Governing the future. Institutional responses to long-term challenges in Sweden and the Netherlands*? Paper to the Council for Europeanists conference, Boston.

Baker, S. (2006). *Sustainable development*. London: Routledge.

Beck, U. (1992). *Risk society: Towards a new modernity*. New Delhi: Sage Bell.

Beland Lindahl, K., & Westholm, E. (2011). Future forests: Perceptions and strategies of key actors. *Scandinavian Journal of Forest Research, 27*(2), 154–163. doi:10.1080/02827581.2011.635073.

Bell, D. (1999). *The coming of post-industrial society: A venture in social forecasting*. New York: Basic Books, first edition, 1973.

Carson, R. I. (1962). *Silent spring*. Boston: Houghton Mifflin.

Ewing, A., Reed, A., Galli, J., Kitzes, J., & Wackernagel, M. (2010). *Calculation methodology for the National Footprint Accounts, 2010 edition*. Oakland: Global Footprint Network.

Giddens, A. (1990). *The consequences of modernity*. Cambridge: Polity Press.

Giddens, A. (1994). Living in a post-modern society. In U. Beck, A. Giddens, & S. Lash (Eds.), *Reflexive modernisation*. Cambridge: Polity Press.

Giddens, A. (1999). Risk and responsibility. *Modern Law Review, 62*(1), 1–10.

Giddens, A. (2009). *The politics of climate change*. Cambridge: Polity Press.

Hägerstrand, T. (1985). Time and culture. In G. Korsch, P. Nijkamp, & K. Zimmerman (Eds.), *The formulation of time preferences in a multidisciplinary perspective: Their consequences for individual behaviour and collective decision-making* (pp. 33–42). Aldershot: Avesbury.

Hölzl, R. (2010). Historicizing sustainability: German scientific forestry in the eighteenth and nineteenth centuries. *Science as Culture, 19*(4), 431–460.

Hornborg, A. (2010). *Myten om maskinen*. Essaer om makt modernitet och miljo. Goteborg: Daidalos. [In Swedish]

Lakoff, A. (2007). Preparing for the next emergency? *Public Culture, 19*(2), 247–271.

Lisberg Jensen, E. (2006). Sätt stopp för sprutet! Från arbetsmiljöproblem till ekologisk risk 1970-talets debatt om hormonslyr och DDT i skogsbruket. In F. Björk, P. Eliasson, & B. Fritzbøger (Eds.), *Miljöhistoria över gränser* (pp. 197–230). Malmö: Malmö högskola. [In Swedish]

Lundqvist, T., & Carlsson, M. (2004). *Framtidspolitiken. Visionen om ett hållbart samhälle i svensk politik*. Stockholm: Institute for Futures Studies.

Meadows et al. (1972). *The limits to growth*. Club of Rome.1972. New York: New Am Library Universe Books.

Niestroy, I. (2005). *Sustaining sustainability* (EEAC series background study No 2). Berlin: European Environment and Sustainable Development Advisory Council.

Our common future (1987). Brundtland Report, United Nations World Commission on Environment and Development (WCED).

Palme, J., Bergmark, Å., Bäckman, O., Estrada, F., Fritzell, J., Lundberg, O., Sjöberg, O., Sommestad, L., & Szebehely, M. (2001). *Welfare in Sweden: The balance sheet for the 1990s*. Stockholm: Fritzes.

Pellizoni, P. (2011). Governing through disorder: Neoliberal environmental governance and social theory. *Global Environmental Change, 21*(3), 795–803.

Governmental Proposition 1992/1993: 226.

Rockström, J. et al. (2009, September 24). A safe operating space for humanity. *Nature, 461*, 472–475. doi:10.1038/461472a. Published online 23 September 2009.

Sardar, Z. (2010). Welcome to postnormal times. *Futures, 42*, 435–444.

Sassken, S., & Dotan, N. (2011). Delegating, not returning, to the biosphere. How to use the multiscalar and ecological footprints of cities. *Global Environmental Change, 21*(3), 823–834. doi:10.1016/gloenvcha. 2011.03.018.

Scoones, I., Leach, M., Smith, A., Stagl, S., Stirling, A., & Thompson J. (2007). *Dynamic systems and the challenge of sustainability* (STEPS Working Paper 1). Brighton: STEPS Centre.

Scott, J. C. (1988). *Seeing like a state: How certain schemes to improve human conditions have failed*. New Haven: Yale University Press.

SEAC. (2007). *Background report to the Swedish Environmental Advisory Council memorandum*. 2007:1. Regeringskansliet

Skovdahl, B. (2012). I: Framtiden som vetenskapligt fält. I *Nycklar till framtiden*. Red: Alm, S., Palme, J., Westholm, E., 2012. Institutet för Framtidsstudier och Dialogos. Stockholm. [In Swedish]

Wästfelt, A. (2006). Minnet av det bortglöma landskapet. I: Jansson, U. (red): *Med landskapet i centrum. Kulturgeografiska perspektiv på nutida historiska landskap*. Meddelande 119/kulturgeografiska institutionen, Stockholms Universitet. [In Swedish]

Westholm, E. (2010). Spatial perspectives on the welfare state. Geography Ymer Yearbook 2010 "*Placing human geography—Sweden through time and space*." Svenska Sällskapet för Geografi och Antropologi.

Wirén, E. (1985). *Allemans skog: slutrapport för projektet: Samhället och skogen* (1st ed.). Stockholm: Liber Publ. [In Swedish]

Chapter 3
Future Forest Trends: Can We Build on Demographically Based Forecasts?

Bo Malmberg

The question addressed in this chapter is: can a simple model that relates income growth to demographic change contribute to an understanding of the challenges that the forestry sector will be facing in the coming decades? Per capita income is interesting because of its impact on the ecological footprints of various societies. The exact relationship between per capita income and energy demand, raw material use, or various types of consumption can shift over time, but this does not change the fact that per capita income growth is a major factor underlying many social and environmental trends. Looking more closely at how income growth correlates with demographic change, therefore, provides a good starting point for discussing future forest trends. An income forecast as such does not constitute a forest scenario but will indicate the important future trends very likely to influence future forest scenarios.

3.1 Introduction

Futures studies is a broad scientific field encompassing many genres and methodologies. In its simplest form, however, a study of the future can be quite straightforward and easy to grasp. Consider that an ordinary map in fact constitutes a future study. The map tells you what you can expect tomorrow if you choose to go to the area shown on the map. Experience tells us that maps are usually quite successful at predicting the future: the roads we have identified on our map will take us to our desired destination, just as an accurate maritime map will allow us to sail without running aground on underwater reefs.

Maps can predict the future because some parts of reality change very slowly (Wegener 2004). Many things will be more or less the same 1 year, 5 years, or even

B. Malmberg (✉)
Department of Human Geography, Stockholm University, Stockholm, Sweden
e-mail: bo.malmberg@humangeo.su.se

E. Westholm et al. (eds.), *The Future Use of Nordic Forests*,
DOI 10.1007/978-3-319-14218-0_3

Photo 3.1 Biomass expansion—a response to demographic change?

25 years from now. This implies that forecasting does not necessarily entail coming up with innovative ideas about new things that will happen in the future. It can be much easier than that. It is enough simply to be able to say what things are likely to stay the same for 10, 20 years, or longer.

The map metaphor is also good for helping us think about why we need futures studies. A good map makes it much easier to plan activities that will be carried out in an environment we have not seen. The same is true of futures studies: we have yet to visit the future, so we need a tool that helps us think rationally about what that future may be like. Then, of course, we need to know how to go about building an understanding of future trends.

Here the map method will help us get the directions right. We need to consider what will *not* change in the future, and use such slowly changing conditions as starting points for outlining the contours of the future landscape. In this chapter, I will argue that these considerations support the idea of using demographic trends as the main tool for constructing future scenarios.

Populations change slowly, and mainly in ways that are easy to predict. Most people who will be alive on this planet 25 years from now are already among us. Very stable mortality patterns allow us to accurately predict how many of those living today will still be alive in 2040. Moreover, as everyone gets one year older for every year they live, we also know the age structure of this surviving population. Population forecasts of the future age compositions of national populations are therefore very likely to be achieved. Hence, the demographic composition of the future is a good starting point, perhaps the best that we have, for looking into the future.

Using population projection for income-growth and environmental forecasting is not a new idea. The traditional approach has mainly been to use population forecasts as an input for calculations of the future labor supply. More recently, population forecasts have also been used as a component in predictions of the future savings rate (Kerdrain et al. 2010). Other key determinants such as education levels and total factor productivity are normally modeled as determined by processes of global convergence. However, income growth forecast can also be based on reduced-form models using only demographic variables as inputs, as there are strong, long-term links between shifts in population structure and trends in per capita income levels (Lindh and Malmberg 2007). Recent research has demonstrated that these relationships are very stable over time (de la Croix et al. 2009; Azomahou and Mishra 2008). Therefore, given that long-term population projections also tend to be relatively accurate, it can be very interesting to explore how the future will look if the historical relationships between demography and income growth also hold up during the next 30–40 years.

Thus, the question addressed in this chapter is: can a simple model that relates income growth to demographic change contribute to an understanding of the challenges that the forestry sector will be facing in the coming decades? The reason why per capita income is interesting is, of course, its impact on the ecological footprints of various societies. The exact relationship between per capita income and energy demand, raw material use, or various types of consumption can shift over time, but this does not change the fact that per capita income growth is a major factor underlying many social and environmental trends. Looking more closely at how income growth correlates with demographic change, therefore, provides a good starting point for discussing future forest trends. An income forecast as such does not constitute a forest scenario but will indicate the important future trends very likely to influence future forest scenarios.

3.2 The State-of-the-Art Approach

Before reviewing the argument for using a simplified demographic model of income growth, it is important to discuss the current state of the art of global income growth forecasting. First, it can be noted here that, generally, long-term income-growth forecasting is not something in which leading economists aspiring to be Nobel Prize winners frequently engage. Instead, most forecasts come from international bodies such as the Organisation for Economic Co-operation and Development (OECD). Environmental concerns have also been an important factor behind the elaboration of growth forecasts, with the report of the Intergovernmental Panel on Climate Change (IPCC) on emission scenarios being one of the first to provide global income growth projections (Nakicenovic 2000).

One possible explanation as to why long-term projections have been rare is that the standard Solow model of economic growth has been seen as unable to explain the persistence of large income gaps between poor and rich countries (Romer 1986). Later, it was clarified that the gaps could be explained by low levels of capital accumulation in poor countries (Mankiv et al. 1992), but the large differences in capital accumulation still needed to be explained. A first attempt to use the Solow model directly to model global growth was made by Wilson and Purushothaman (2003). They show in their paper that the convergence predicted by a Solow model might generate very high growth rates for countries such as Brazil, Russia, India, and China (the "BRIC" countries). The title of their paper "Dreaming with the BRICs," however, shows that the authors were hesitant to trust their own results. Since then, confidence that Solow-type convergence across a wide range of countries is a real possibility has grown in the light of the growth performance in China, India, and Latin America. Thus, recently, the OECD and others have published long-term growth forecasts that are essentially adaptations of the Solow model (Duval and de la Maisonneuve 2010; Fouré et al. 2012; Johansson et al. 2013). To generate a growth forecast, these models use projections of labor supply, education levels, capital accumulation, and total factor productivity that are based on population forecasts and convergence assumptions. Nevertheless, there seems to be some hesitation about accepting the outcome of the forecast. Thus, Duval and de la Maisonneuve (p. 15) state:

> Long-run growth projections are inherently speculative due to model and parameter uncertainty, but also because the various determinants of growth would remain very hard to predict even in a hypothetical case where the "true" model were known.

A possible reason for this reluctance to stand by their own forecast could be that, historically, convergence in total factor productivity and rates of capital accumulation have not always been the rule. Despite being based on the best methods available, these forecast models lack one important justification. It has not been demonstrated that the convergence assumption they are built on has been able to explain the historical growth patterns of developing countries. It can be argued that this lack of belief in the forecasts can limit their value for researchers and policymakers. If one is to act on the basis of a forecast, it is not enough for it to be up to standard technically. One must also believe that the future will in fact evolve in the way the forecast predicts—or at least with a high degree of probability. And here an analysis of the historical relationships between demographic change and income growth becomes important. If it can be demonstrated that a trend toward global income convergence is not the result of a potentially arbitrary assumption but is what should be expected if historically given relationships remain unchanged, this should increase the willingness to use long-term income forecast as a guide for future action.

3.3 The Demographic Dividend

A simple argument for using, for example, the population projections by age and sex published by the Population Division of the United Nations as a basis for scenarios of future trends is that the numbers contained in those tables do tell us, with a comparatively high degree of certainty, what the future will look like. However, the rows and columns of numbers that make up those projections do not, on their own, tell a very interesting story. It is only through the research results from economics and other disciplines over the last 25 years that those projections can tell us stories of what the future will be like.

Starting in the late 1980s and early 1990s, a number of scholars began to look into the question of how age structure change affects social and economic development (Kelley and Schmidt 1996; Crenshaw et al. 1997; Malmberg and Sommestad 2000; Bloom et al. 2003; de la Croix et al. 2009). One possible reason for this research starting at that time is that the Population Division began to make age structure data for the post-war period available electronically (first on floppy disks). This gave researchers a new rich dataset, and the reward was a series of important findings. It turned out that population age structure is strongly correlated with a series of socioeconomic variables, for example, per capita income, income growth, savings and investment rates, public expenditure, inflation, current account deficits, and violent social conflict. Given these strong correlations, predicting future trends in these variables becomes very straightforward. The only thing needed is to link the age effects estimated using historical data with UN or other population projections and the result is a projection of the variable of interest.

Is it possible that this mechanistic way of conducting a future study can really work? One important theoretical argument as to why this approach is sound is that the effects of age structure on macro variables are very likely linked to the dramatic age-related shifts in economic and social behavior that characterize the human life cycle. People aged 0, 1, 4, and 8 function in very different ways; this is also true of people aged 12, 16, 24, and 32. Age-related changes also continue in adulthood, with people aged 42, 54, 66, and 78 experiencing changes in career, economic status, family situation, and health. Moreover, although life-course patterns change on the margins, the general patterns of childhood, adolescence, early adulthood, family formation, and retirement remain the same. Migration, for example, is mainly concentrated in the 18–30 age group today, just as it was 150 years ago. Thus this stability in life-course patterns is a reason for using population forecasting as a tool in looking for possible future trends.

In addition to the above arguments, there is one more reason for using demography as a guide to future socioeconomic trends. This is the fact that demographic change over the last 200 years has not been an erratic process of trends shifting back and forth. On the contrary, the most important demographic trends have remained remarkably uniform in their patterns, with the timing of various trends being the key difference between countries (Dyson 2010). Population researchers call this macro trend the demographic transition, a long-term process by which birth and death rates fall from high to low levels. This process tends to start with a decline in death rates, whereas the start of the birth rate decline is typically delayed some 30–60 years after the start of the mortality decline. The first demographic transitions occurred in northwestern Europe and in the United States (USA). The same process later also occurred in the rest of Europe and in Japan. Since World War II, the demographic transition has affected the rest of the world, with Asia and Latin America being ahead of sub-Saharan Africa. The concept of a demographic transition has been around since 1945, although the general validity of the theory was questioned in the 1960s. However, in view of a generalized fertility decline in most parts of the world except sub-Saharan Africa over the last quarter of the twentieth century, the existence and near-universality of the demographic transition process has become an acknowledged fact.

The importance of the demographic transition for social and economic development is closely linked to its effect on the age structure (Malmberg and Sommestad 2000). In the early phases of the transition, mortality declines but fertility stays constant. This results in a rapid expansion of the child population and a rapid increase in the child dependency rate. This expansion occurs because the mortality decline at that point leads to a dramatic decline in infant and child mortality; this increases the number of survivors, and when they reach fertile ages, the number of births also increases. When fertility declines, this pattern changes dramatically: the increase in the child population stops, and population growth will instead be concentrated mainly in the working ages. This is a result of the maturing of the large cohorts born during the high-fertility, low-mortality phase, and it leads to a drastic reduction in the child dependency rate. The final phase of this age transition occurs when these large cohorts go into retirement and the smaller cohorts born during the low-fertility phase enter working ages. At that point, the dependency rate again increases, but now it is the old-age dependency rate that increases. The increasing share of the old-age population may also be more pronounced because of continuing declines in old-age mortality and because of fertility levels that continue to decline to below two children per women (Dyson 2010).

Over the last 20 years, researchers have be able to demonstrate that patterns of social and economic development are closely linked to the age structure shifts that occur during the demographic transition. The increase in child dependency that occurs during the first phase tends to be associated with poor economic performance and increasing poverty. With the decline in child dependency that occurs after the fertility decline, the economic effect of the age transition becomes positive (Bloom et al. 2003). Countries in this phase tend to demonstrate very high growth rates in GDP per capita. Japan 1950–1990 is one example; Korea, Taiwan, Hong Kong, and Singapore 1960–2000 are others, and China 1970–2010 and India 1990-the present are more recent examples. European examples include Sweden 1880–1940, Spain 1960–2000, and Ireland 1970–2010. This effect has been called the demographic dividend, and it is now acknowledged as a key factor in generating economic development.

On the global level, this age transition can explain why the first decades after World War II did not lead to a global convergence of per capita income but instead to widening global income gaps. The reason is that child dependency rates rose rapidly in the developing countries because of a large reduction in infant mortality (Davis 1956), whereas the already rich countries could benefit from a maturing working age population (Lindh and Malmberg 1999). It is only with declining fertility after 1970 that the developing countries have begun to experience declining dependency rates, increasing growth rates of the working-age population, and an acceleration of per capita income growth. Thus, an important contribution of the expanding literature on the effects of the demographic transition on economic development is that it explains why the Solow growth model for many decades did not work very well when applied to the developing countries. But this literature also explains why Solow convergence is much more likely from 1990s and onwards.

3.4 Forecasting Per Capita Income Growth in the Twenty-First Century

One way to demonstrate the importance of demographic shocks for economic development is to estimate a highly simplified model that uses only demographic variables to account for yearly shifts in per capita income. The model uses data for 111 countries from 1961 to 1996. The demographic data comes from the Population Division dataset (United Nations 2005) and the GDP data from the Penn database (Heston et al. 2006). In the model, per capita income (in logs) is regressed on population shares in the 0–14, 15–29, 30–49, 50–64, and 65+ age groups (also in logs). Acknowledging that health is a major determinant of economic development alongside age structure, the model also includes log life expectancy interacted with the age variables. This model specification allows the age parameters to shift with any large shifts in life expectancy. Theoretically, this can be justified by results indicating that decisions about education and saving are influenced by life expectancy. It also produces a model that generates very stable results.

Table 3.1 Parameter estimates for the dividend model

	Level of life expectancy				
	40	50	60	70	80
Intercept	19.96	21.17	22.16	22.99	23.72
Log 0–14	−1.53	−1.30	−1.10	−0.94	−0.80
Log 15–39	0.49	0.29	0.13	0.00	−0.12
Log 30–49	0.73	0.55	0.40	0.27	0.16
Log 50–64	0.31	0.32	0.32	0.32	0.33
Log 65+	−0.69	−0.27	0.07	0.36	0.61

The results of this estimation are shown in Table 3.1. The estimates indicate that as long as life expectancy is 60 years and lower, age effects on per capita income are hump-shaped. That is, the population shares of working age groups have a more positive effect than do the shares of children or old-age adults. As life expectancy increases, the effect of older age groups dominates the effect of younger age groups.

Will this model provide good predictions? According to forecasting experts, the best way to answer this question is to look at out-of-sample performance, as the traditional R^2 has been demonstrated to be a very poor guide to forecasting performance. Formal out-of-sample tests provided in Lindh and Malmberg (2007) indicate that the model is stable. Here, an alternative based on de la Croix et al. (2009) is presented. As it is difficult to test model performance in the forecasting period, an alternative is to use the model for backcasting: can the model reproduce the pattern of per capita income growth that a country has experienced before the period used in the estimation? Such backcasting is possible for Sweden, as we have data on age structure, life expectancy, and gross domestic product (GDP) extending back into the nineteenth century.

The results are shown in Fig. 3.1. Here, the red line indicates the per capita income level in Sweden predicted by the model, estimated using modern global data. As can be seen, the model works well at least back to 1870. This indicates that the relationships between demography and GDP per capita captured in the estimated model are highly persistent. Growth in GDP per capita seems to go in tandem with the demographic changes that occur during the demographic transition.

3.5 Per Capita Income Forecast, 2005–2050

The impressive performance of this simple demographic model in out-of-sample tests and in a backcasting experiment suggests that it can also be useful for predicting future income growth. Figure 3.2 shows the results of such a forecast of per capita income forecasts for 18 regions in the world. The forecasts are based on the model presented in Table 3.1.

The top six panels show the growth trends in today's developed countries. These forecasts indicate that growth will continue but at somewhat declining rates,

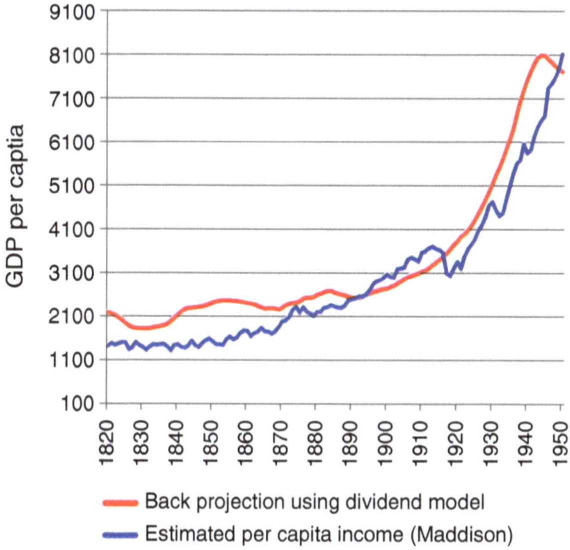

Fig. 3.1 GDP per capita in Sweden 1820–1950. Maddison data and back projection based on demographic dividend model (de la Croix et al. 2009)

especially in southern Europe. Per capita income is expected to increase to about USD 50,000 per year in west, southern, and northern Europe and to about USD 70,000 in North America and Oceania.

In eastern, southeastern, and western Asia, per capita income is expected to rise to somewhat below USD 30,000, and in south-central Asia to somewhat below USD 20,000. In comparison, per capita income in northern Europe reached USD 30,000 in 2000 and USD 20,000 in 1980.

In Latin America and Central America, even stronger growth is expected. Here per capita income is projected to increase to about USD 40,000. However, in the Caribbean, somewhat weaker growth is expected, per capita income increasing to about USD 20,000.

In Africa, strong growth is expected in northern Africa, with per capita income approaching USD 30,000 in 2050. In other parts of Africa, the growth prospects are weaker. Middle and southern Africa are expected to reach about USD 8,000 in per capita income, but eastern Africa will only reach about USD 4,000 in per capita income. The main reason for this poor performance is the continuing high fertility and slow decline in mortality.

The growth rates implied in the above projection are approximately 3 % per year for high-growth countries, close to 1 % for high-income countries, and approximately 2 % per year for the rest. The per capita income will double, and double again in Central America, Latin America, south-central Asia, and northern Africa according to this projection.

Clearly, these increases in per capita income will bring about dramatic changes in the consumption patterns of the countries involved. If historical patterns can be

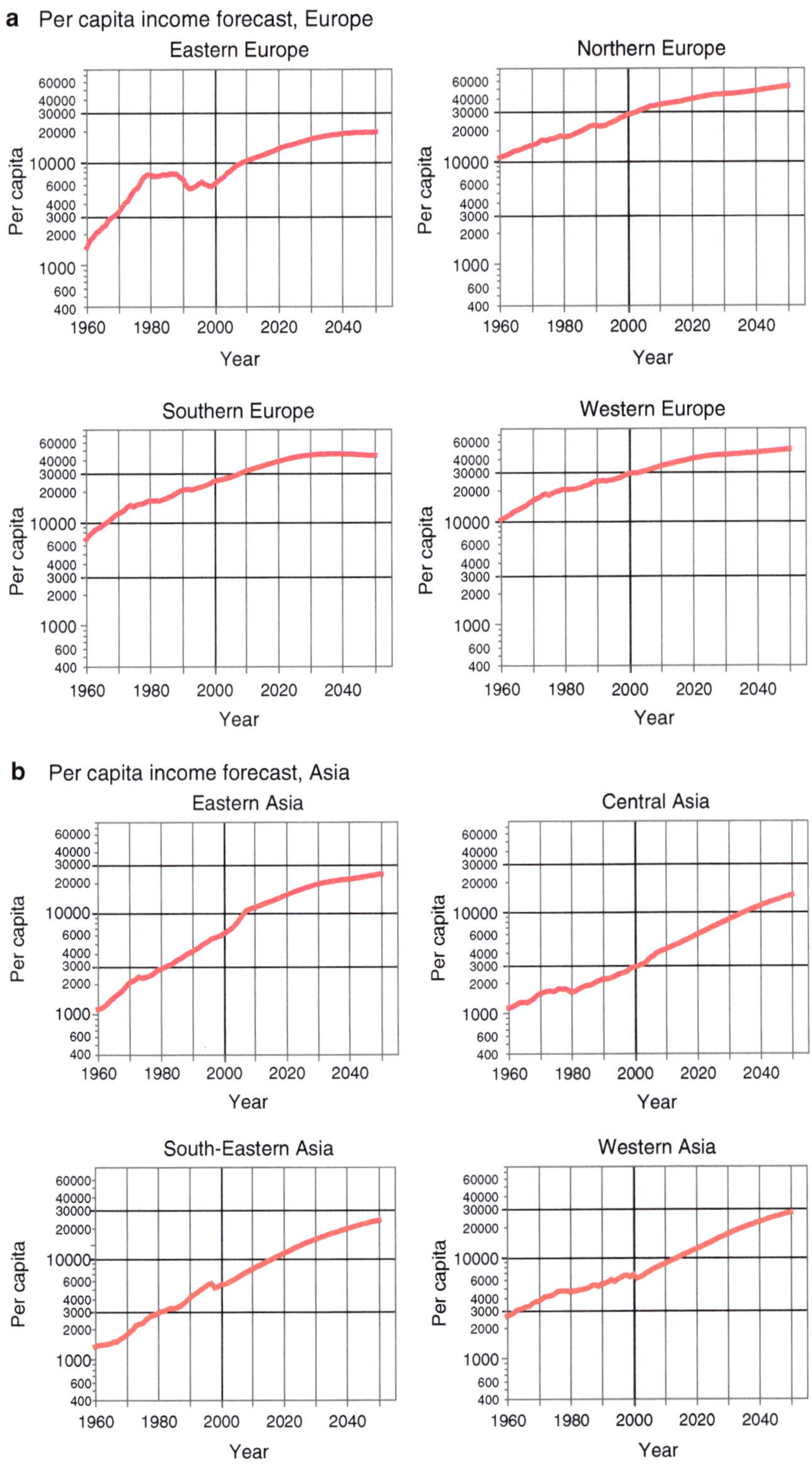

Fig. 3.2 Per capita income forecast, 18 world regions

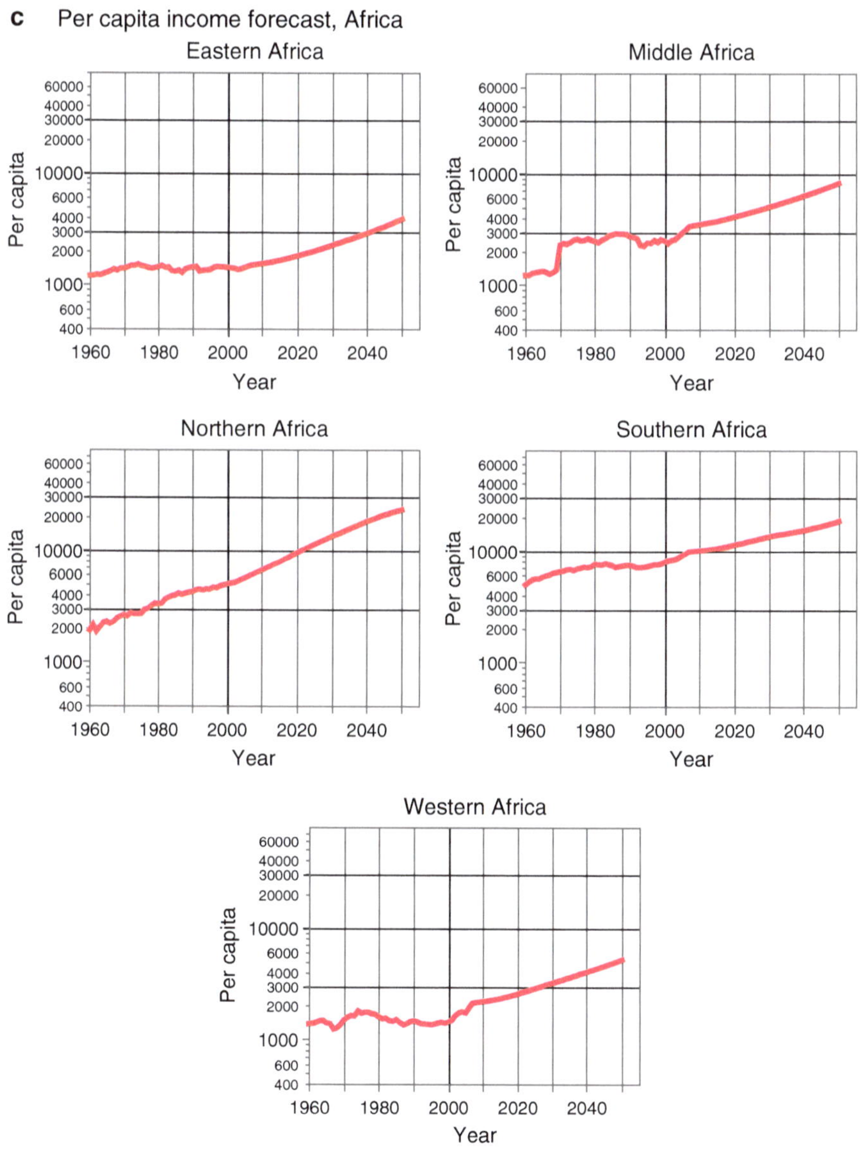

Fig. 3.2 (continued)

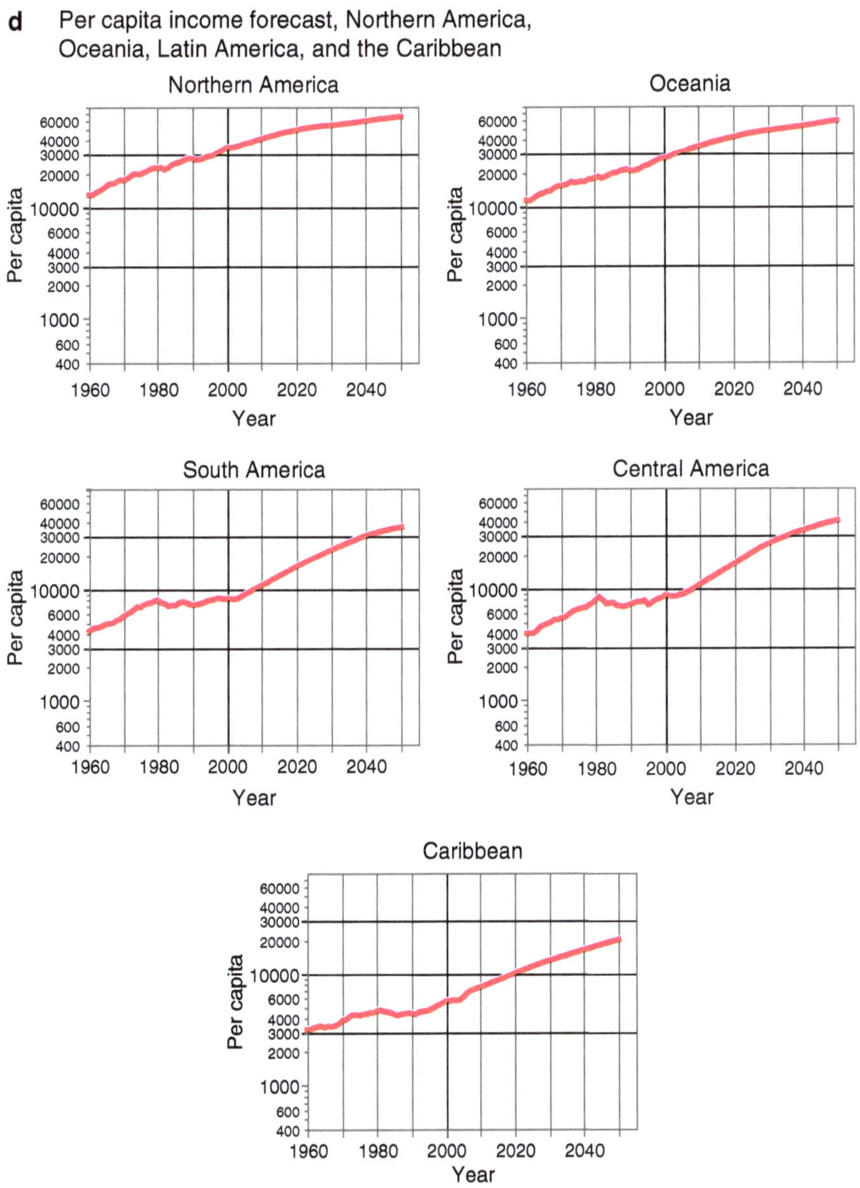

Fig. 3.2 (continued)

taken as a guide, per capita income growth from around USD 5,000 per year to above USD 20,000 per year tends to generate demand for goods that require both high energy and high material inputs. Environmental pressures can therefore be expected to be high. This is even more so if one takes into account the population size of the countries that will experience rapid income growth.

3.6 GDP Forecast, 2005–2050

Table 3.2 presents the 2050 forecast for total gross domestic product (GDP) along with population figures and GDP per capita for 18 regions of the world. The corresponding values for 2005 are also presented to enable comparison.

The results presented in Table 3.2 indicate that total world GDP can be expected to more than triple between 2005 and 2050, mostly due to an increase in per capita income (more than a doubling), with population growth playing a smaller role. The total increase in annual GDP will be USD 142 trillion, a figure that is so large that it is difficult to make sense of. One possible comparison is with the current GDP of Sweden, which is about USD 0.275 trillion. This implies that the GDP added between 2005 and 2050 is comparable to adding 500 countries with the same consumption level and population as Sweden, in other words, adding approximately 4.7 billion people with the same consumption level as that of today's Sweden.

Duval and Maisonneuve (2010) in their paper do not use the same aggregates as in Table 3.2, which makes the result difficult to compare in detail. First, it can be noted that these authors predict a growth rate for world GDP that is 3.5 % for the 2005–2050 period, whereas the prediction in Table 3.2 is 2.6 %. An important source of this difference could be the Indian forecast where the authors predict a growth rate of 6.2, whereas Table 3.2 gives a growth rate for South Asia of 4.1 %. For other regions the differences are smaller. The general growth pattern suggested by the OECD and by the demographic model, thus, both point in the same direction.

It should be noted that one assumption underlying these income forecasts is that historically observed links between demographic structure and per capita income are not made invalid by, for example, dramatic shifts in climate. On the other hand, the available estimates of the effects of climate change on economic growth indicate that the magnitude of these effects, although economically important, will not become overwhelming as long as the temperature increase is moderate (Stern 2007). Given the margin of error of the above forecast, I would conclude that they remain valid for climate scenarios in which the temperature increase stays below 3 °C.

Table 3.2 Global change per region, 2005–2050: Population, GDP, and GDP per capita

Region	Population, 2005, millions	Population, 2050, millions	GDP 2005, billion dollars	GDP 2050, billion dollars	Change in GDP, trillion dollars	Index 2050, 2005 = 100	Per capita, 2005	Per capita, 2050
Eastern Africa	234	565	337	2,215	1.878	657	1,442	3,923
Middle Africa	53	124	158	1,033	0.875	654	2,961	8,353
Northern Africa	150	234	857	5,504	4.647	642	5,701	23,477
Southern Africa	54	66	494	1,235	0.741	250	9,156	18,814
Western Africa	267	617	468	3,256	2.787	695	1,757	5,279
Eastern Asia	1,440	1,519	12,303	37,293	24.990	303	8,545	24,556
South-Central Asia	1,567	2,339	5,506	34,406	28.900	625	3,514	14,708
South-Eastern Asia	401	553	2,576	13,253	10.677	514	6,431	23,987
Western Asia	103	156	753	4,320	3.567	574	7,280	27,627
Eastern Europe	22	17	178	341	0.163	192	8,211	19,731
Northern Europe	89	107	2,818	5,641	2.824	200	31,621	52,931
Southern Europe	123	129	3,373	5,799	2.426	172	27,355	44,859
Western Europe	104	114	3,191	5,707	2.516	179	30,720	49,924
Caribbean	23	33	148	691	0.544	469	6,363	20,869
Central America	144	196	1,322	8,301	6.979	628	9,181	42,280
South America	370	481	3,410	18,065	14.655	530	9,212	37,542
Northern America	605	808	23,148	54,633	31.485	236	38,230	67,627
Oceania	25	35	806	2,125	1.319	264	31,812	60,754
Total 108 countries	**5,775**	**8,093**	**61,846**	**203,821**	**142**	**330**	**10,709**	**25,185**

3.7 Consequences for Consumption and Resource Use

Clearly, the increase in GDP will greatly challenge global production and resources. For example, in Sweden there is about one car for every two Swedes (Statistics Sweden 2012). Adding 500 Sweden would therefore imply another 2.4 billion cars in the world. The consequences of such an expansion would be vast. First, the demand for raw material would increase greatly. Second, these vehicles would need to be fueled, which implies a large increase in energy demand. Third, the expanding demand would increase the return on investment in new car technology, implying that the energy and raw material demand will not increase in proportion to the increase in the number of cars. On the contrary, major breakthroughs in construction and fuel technology should be considered real possibilities. It is fully possible that, in the absence of such breakthroughs, the projected increase in GDP will not materialize. That is, if expanding demand comes up against resource supply barriers that cannot be circumvented by new technology, this is likely to put a brake on increases in real GDP per capita, much as Malthus predicted some 200 years ago. On the other hand, a fixed supply of natural resources is nothing new. Natural resources were also fixed during the period for which our model was estimated, 1950–2000. What the model suggests is that, over the last 50 years, technological development has been endogenous to the demographically conditioned growth process. This, in my view, makes it possible to argue that we can expect endogenous technological change in the coming decade that is at least as fast as it has been up to now.

Another example: per capita paper consumption in Sweden is approximately 250 kg per person (Sweden Forest Industries Federation 2010). Current per capita consumption in the world has been estimated at approximately 50 kg, corresponding to global production of approximately 300 million tons. Adding 500 Sweden would thus correspond to an increase in world paper production of 1,000 million tons. In fact, the results of this calculation are very close to those obtained by Hedenus and Johansson (2007) in a report to the Swedish government. Hedenus and Johansson combine the income forecast of Lindh and Malmberg (2007) with a model that relates per capita consumption of paper and paperboard to per capita income. Their results indicate an increase in consumption from approximately 300 million tons to 1,200 million tons over the 2000–2050 period, most coming from China and other non-OECD countries. For sawnwood and wood-based panels, the authors forecast an increase from under 600 million m^3 to over 1,200 million m^3. With paper recovery taken into account, they estimate that that the total demand for industrial roundwood will increase from 1,500 m^3 to 3,500 m^3. Hedenus and Johansson point out that these estimates are considerably higher than other estimates. Moreover, this is a demand-based estimate that does not consider supply conditions. Still, the analysis suggests that demand factors are likely to push the forestry sector toward more intensive production methods with the aim of increasing output.

Today, Sweden's primary energy supply is approximately 2 exajoules (1 EJ = 10^{18} J). Adding 500 Sweden would therefore imply an increase in world energy demand to 1,000 exajoules from today's 400. This can be compared to Hedenus and Johansson's estimated increase in primary energy demand from 400 to 1,200 exajoules (1 EJ = 10^{18} J), that is, an 800-exajoule increase.

These two examples clearly indicate that the per capita income growth which can be expected as a result of the completion of the demographic transition is likely to yield a very large increase in demand for forest resources in the form of wood products. Moreover, the increase in energy demand could make forest-based energy production another important source of demand for forest resources.

The Hedenus and Johansson forecast includes the decoupling of income growth and resource demand to the extent that such a tendency is present in the data they use to estimate their resource-demand income growth relationships. Thus, the forecast is conditioned on a historical relationship that could change in response to technological shifts. The Hedenus and Johansson forecast also differs somewhat from forecasts made using other approaches (see Nilsson, Chap. 4, this book). Experience indicates, however, that the best forecasting results are obtained when forecasts made using different approaches are combined. The trend presented above should therefore be seen as a contribution to the pool of forecasts that, taken as a whole, can help us map an uncertain future.

3.8 Conclusions

In his book *Principles of forecasting*, Armstrong (2001) argues that predictions can be improved by combing forecasts that uses different methods. In this paper the use of demographically based long-term forecasts of global income growth as a complement to state-of-the art global forecasting models such as those produced by the OECD has been discussed. My conclusion is that population-based forecasting can play an important role. First, predictions can be improved by combining forecasts that use different methods (Armstrong 2001). If results arrived at using different approaches point in the same direction our confidence in the numbers will be increased. And, as I have argued above, if one has no confidence in a forecast it will be of very limited value. Second, the results obtained using the demographic approach are also important because they show that long-term forecasting of strong but uneven global income growth is what can be expected if historical correlation between demographic shifts remains stable for the next 30–40 years. This argument cannot be made with the same force for the state-of-the-art model. Finally, the demographic model provides an accessible explanation as to why we should expect a particular pattern of global income growth in the coming decades in that it links the forecasts to the research results from the last 20 years which have demonstrated that demographic change plays a key role for economic development.

Above, I have also briefly discussed the implications of income growth for future forest trends. However, the chapter has not sought to go more deeply into this question. Instead, I wished to point out that certain global trends can be predicted with relative certainty and that the consequences of this deserve to be explored in more detail. One area that will be of interest here is how pressures generated by increasing global income will affect forest management practices in the coming decade.

Here, two possibly conflicting trends can be seen. In the short run, increasing demand can lead to the depletion of forest resources. In the longer run, though, increasing demand for forest services is likely to stimulate management practices that increase the long-term productivity of forest lands. More intensive forest management practices are, in my opinion, the most likely outcome of the changing global situation.

A probable outcome of this is that the global trend toward deforestation observed in recent decades is likely to reverse. This fits with a historical pattern whereby in many countries long-term deforestation trends have been followed by afforestation. Although this forest transition theory is still not completely developed or accepted, afforestation is a not unlikely scenario for the coming decades. What should be noted here is that the forest transition is not simply a spontaneous market-led process. Instead, in many cases, reforestation is promoted by changing perceptions of the need for sustainable forest practices and by policy measures that that favor the conservation of forest resources. It is also important to acknowledge that afforestation will not bring back the more or less pristine forests that were cut down during the deforestation phase. Instead, the concern for forest output is likely to stimulate the establishment of highly managed forests that differ considerably from old-growth forests.

References

Armstrong, J. S. (2001). *Principles of forecasting: A handbook for researchers and practitioners* (Vol. 30). Dordrecht: Springer.

Azomahou, T., & Mishra, T. (2008). Age dynamics and economic growth: Revisiting the nexus in a nonparametric setting. *Economics Letters, 99*(1), 67–71.

Bloom, D. E., et al. (2003). *The demographic dividend: A new perspective on the economic consequences of population change*. Santa Monica: Rand.

Crenshaw, E. M., Ameen, A. Z., et al. (1997). Population dynamics and economic development: Age-specific population growth rates and economic growth in developing countries, 1965 to 1990. *American Sociological Review, 62*(6), 974–984.

Davis, K. (1956). The amazing decline of mortality in underdeveloped areas. *The American Economic Review, 46*, 305–318.

de la Croix, D., et al. (2009). Demographic change and economic growth in Sweden: 1750–2050. *Journal of Macroeconomics, 31*(1), 132–148.

Duval, R., & de la Maisonneuve, C. (2010). Long-run growth scenarios for the world economy. *Journal of Policy Modeling, 32*(1), 64–80.

Dyson, T. (2010). *Population and development: The demographic transition*. London: Zed.

Fouré, J., Bénassy-Quéré, A., & Fontagné, L. (2012). *The great shift: Macroeconomic projections for the world economy at the 2050 horizon*. Working paper of the *Centre d'études prospectives et d'informations internationales*. http://www.cepii.fr/PDF_PUB/wp/2012/wp2012-03.pdf. Accessed 13 Jan 2014.

Hedenus, F., & Johansson, D. J. A. (2007). Global scenarios for energy, carbon dioxide and forest products. In *Scenarios on economic growth and resource demand. Background report to the Swedish Environmental Advisory Council Memorandum 2007:1 Miljövårdsberedningen*. Stockholm: Swedish Environmental Advisory Council.

Heston, A., et al. (2006). *Penn world table version 6.2*. Philadelphia: Center for International Comparisons at the University of Pennsylvania (CICUP).
Johansson, Å., Guillemette, Y., Murtin, F., Turner, D., Nicoletti, G., de la Maisonneuve, C., Bagnoli, P., Bousquet, G., & Spinelli, F. (2013). *Long-term growth scenarios* (No. 1000). Paris: OECD Publishing.
Kelley, A. C., & Schmidt, R. M. (1996). Saving, dependency and development. *Journal of Population Economics, 9*(4), 365–386.
Kerdrain, C., Koske, I., & Wanner, I. (2010). *The impact of structural policies on saving, investment and current accounts*. (OECD Economics Department Working Papers, No. 815). Paris: OECD Publishing.
Lindh, T., & Malmberg, B. (1999). Age structure effects and growth in the OECD, 1950–1990. *Journal of Population Economics, 12*(3), 431–449.
Lindh, T., & Malmberg, B. (2007). Demographically based global income forecasts up to the year 2050. *International Journal of Forecasting, 23*(4), 553–567.
Malmberg, B., & Sommestad, L. (2000). The hidden pulse of history. Age transition and economic change in Sweden, 1820–2000. *Scandinavian Journal of History, 25*(1–2), 130–146.
Mankiw, N. G., Romer, D., & Weil, D. N. (1992). A contribution to the empirics of economic growth. *The Quarterly Journal of Economics, 107*(2), 407–437.
Nakicenovic, N., & Intergovernmental Panel on Climate Change. (2000). *Special report on emissions scenarios*. Cambridge: Cambridge University Press.
Romer, P. M. (1986). Increasing returns and long-run growth. *The Journal of Political Economy, 94*, 1002–1037.
Statistics Sweden. (2012). *Fordonsstatistik*. From www.trafa.se/Statistik/Vagtrafik/Fordon/
Stern, N. N. H. (Ed.). (2007). *The economics of climate change: The Stern review*. Cambridge: Cambridge University Press.
Sweden Forest Industries Federation. (2010). *Paper consumption in Sweden*. from http://www.forestindustries.se/web/Paper_consumption_in_Sweden_1.aspx
United Nations. (2005). *World population prospects: The 2004 revision*. New York: United Nations, Population Division.
Wegener, M. (2004). Overview of land-use transport models. In D. A. Hensher & K. Button (Eds.), *Transport geography and spatial systems* (Handbook 5 of the handbook in transport, pp. 127–146). Kidlington: Pergamon/Elsevier Science.
Wilson, D., & Purushothaman, R. (2003). *Dreaming with BRICs: The path to 2050* (Global economics paper). New York: Goldman Sachs.

Chapter 4
Global Trends and Possible Future Land Use

Sten Nilsson

This chapter explores the future global need for land by investigating four trends that drive global land use change: future energy demand; future food demand; future demand for various forest products; and climate change. These trends affect land use competition, food prices, and deforestation rates, and they interact with each other in complex ways that are difficult to foresee. In this chapter, a number of trends, their interactions, and possible implications are discussed on the basis of available scenarios and estimates. The chapter ends with a discussion of how global land use changes might influence the conditions for future Nordic forest management and forest sector transition.

4.1 Future Energy Demand and Supply

As discussed in the introduction to this book (Chap. 1), a growing world population with increasing per capita incomes and consumption levels will boost future energy demand. Thus, analyzing demand under different assumptions is of utmost importance. The future global demand for energy has been analyzed by many scenarios over time. A number of more recent and, for this chapter, relevant scenarios present a total energy consumption that varies between 400 and 1,200 EJ/year in 2050, as illustrated by Fig. 4.1 (Shell 2008; Ecofys and WWF 2010; Greenpeace 2010; Shell 2011; IEA 2011; Riahi et al. 2012). The wide range is due to differences in the

---

S. Nilsson (✉)
The Royal Swedish Academy of Agriculture and Forestry,
Stockholm, Sweden

Ecosystems Services and Management Program (ESM), International Institute
for Applied Systems Analysis (IIASA), Laxenburg, Austria
e-mail: stenbnilsson@gmail.com; nilsson@iiasa.ac.at

E. Westholm et al. (eds.), *The Future Use of Nordic Forests*,
DOI 10.1007/978-3-319-14218-0_4

Photo 4.1 Land use pressure is expected

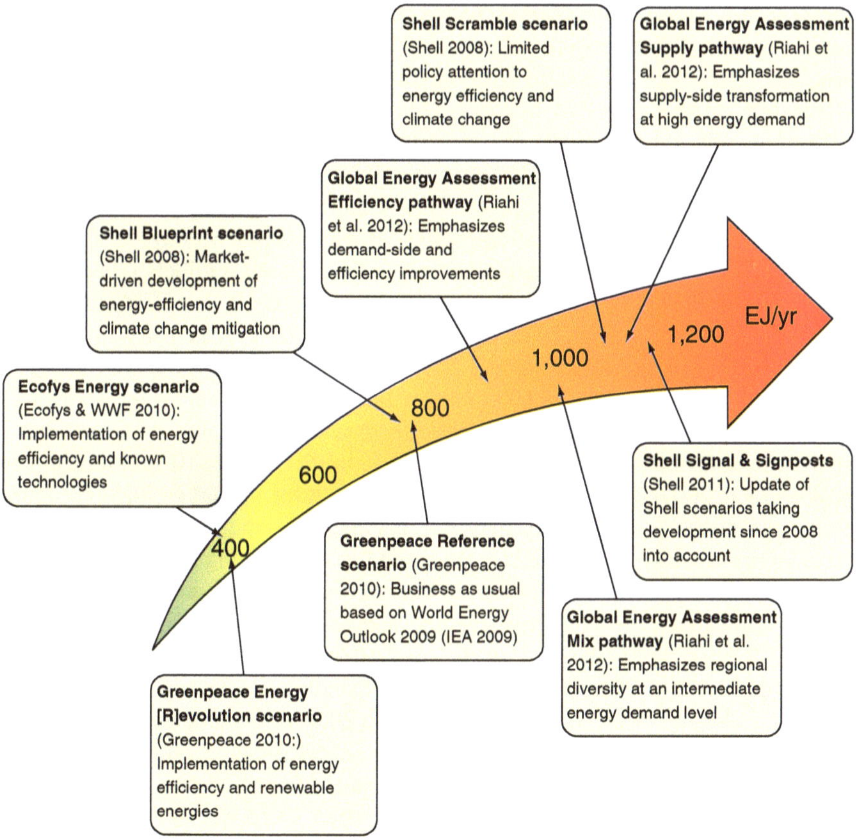

Fig. 4.1 Total energy demand in 2050, as projected by nine different scenarios. The range is wide, between 400 and 1,200 EJ/year, depending on different assumptions concerning future development

underlying assumptions in the models used and in the scenarios with respect to, for example, economic development, energy efficiency, and climate change effects and mitigation measures. For comparative purposes, energy consumption was assessed to be 417 EJ/year in 2000.

For the future of Nordic forestry, assessments of bioenergy demand are of particular importance, and some energy scenarios have paid special attention to the development of biomass-based energy, that is, energy derived from biological sources like agricultural crops, wood, algae, waste, or residues. Global bioenergy use was assessed to be 45 EJ/year in 2005, and the assessment for 2050 is in the range of 130–175 EJ/year (WBGU 2010; Shell 2011; Global Energy Assessment 2012). This corresponds to a need for 1.5–2.5 billion m^3 of biomass, which would potentially necessitate an additional 230–385 million hectares (ha) of land (if in the northern hemisphere) or 50–85 million ha of land (if in the southern hemisphere).

Liquid biofuels are likely to be an important component of future bioenergy, and the consumption of liquid biofuels is thus expected to rise. However, we need to distinguish between the different types of biofuels, as first-generation biofuels are mainly produced from crop-based biomass, for example, sugarcane ethanol or rapeseed biodiesel, while second-generation biofuels are wood-based (ligno-cellulosic). Depending on the source of biomass, biofuels have different direct and indirect effects, not only on land use change, but also on greenhouse gas (GHG) emissions. The relationships are complex, but recent studies indicate that increased bioenergy use would lead to less indirect land use change and fewer GHG emissions if it were based mainly on second-generation biofuels produced from woody biomass from sustainably managed forests rather than on first-generation biofuels (Havlik et al. 2011). The Food and Agriculture Organization (FAO; FAO 2009) makes a forecast of total biofuel consumption to 2018, and when this forecast is recalculated for a longer time horizon, ethanol consumption in 2030 is projected to be 375,000 million liters and biodiesel consumption 150,000 million liters, corresponding to approximately 8.0 EJ and 5.2 EJ, respectively. If we look at first-generation biofuels, the total use of coarse grains for ethanol production reached 110 million tons or 10 % of total global production in 2007–2008. If the consumption of grain for ethanol production continued to constitute 10 % of total grain production in 2030, some additional 200 million ha of cropland would be required for grain and first-generation biofuel production in 2030. If we look at second-generation biofuels, the Food and Agriculture Organization (FAO; Whiteman 2010) has assessed the additional (compared to today) wood required for heat and power and second-generation liquid biofuels in 2030 to be in the range of 1,100–1,700 million m^3/year, corresponding to approximately 12–18 EJ/year. An extension of the Whiteman scenario to 2050 shows a demand for an additional 2,400 million m^3/year (corresponding to approximately 26 EJ/year) at that point in time, a result similar to those in Shell (2011) and the Global Energy Assessment (2012) (see also Kraxner and Nordström, Chap. 5, this book).

Future energy demand, availability, and prices have many direct as well as indirect impacts on the land use sector. Increasing prices of fossil energy will, for example, affect food production via input factors for agriculture such as fertilizers, pesticides, direct energy input, and irrigation. Higher energy prices will make it more profitable to produce both biofuels and other types of bioenergy. Increased bioenergy production may, in turn, lead to increased land costs, impacting food production and food prices. It may also intensify the competition for land and affect whether conventional forestry will be sustained at the current level or expand. However, increasing fossil energy prices will not only cause a drive toward bioenergy; it will also generally stimulate production based on biomass as raw material, namely, a "bio-economy" (Langeveld et al. 2010). These bioproducts may become a substitute for products based on fossil fuels or raw materials which, being tied to fossil energy prices, become too expensive. The production of these bioproducts will further increase the demand for biomass and the competition for land.

A detailed analysis of the energy scenarios leads us to the following overall conclusions:

- *Without strong and urgent measures for transformation of the global energy system, it will be difficult to secure the energy supply (and sustainable development) by 2050*

- *There are great opportunities for transforming the existing energy system through energy efficiency and renewable energies, but the higher the energy demand is in 2050 the more difficult it will be to find sustainable energy solutions*
- *There are concerns that, because of developments in international governance, geopolitics, economy, and urbanization, and the long lead times needed for implementation of new technologies, global society will not be able to transform the global energy system fast enough and that primary energy use may end up in the magnitude of 1,000–1,200 EJ/year in 2050, instead of a possible 400–600 EJ/year*
- *Thus, the lack of formulation and implementation of the required policies and measures is one of the key bottlenecks in the global energy system transformation*
- *The scenario sets are rather consistent with respect to the future extent of biomass-based energies and correspond to 130–175 EJ/year in 2050; this, in turn, corresponds to 1.5–2.5 billion* m^3 *of biomass which, in turn, corresponds to a (potentially additional) requirement for land of 230–385 million ha (if in the northern hemisphere) or 50–85 million ha (if in the southern hemisphere).*

4.2 Future Food Demand and Supply

Most studies about future agriculture and food outlooks have for some time made reference to the 2009 FAO assessment, "How to feed the world in 2050" (Alexandratos 2009; Bruinsma 2009), which anticipates that world farm output will have to rise by some 70 % to feed the population by 2050 (an additional 1,000 million tons of cereals and 200 million tons of meat). This increase is divided between an increase in crop production of 66 % and in meat production of 85 % at a global scale. The total production increase is 97 % for developing countries and 23 % for developed countries by 2050. FAO has the optimistic view that the increased demand can be met by increased supply, if necessary investments and efficient agriculture policies are put in place. In developing countries 80 % of the production increase is assessed to come from increased yield and cropping. This will be a challenge, given that the growth rates of yields of major crops have been declining: from 3.2 % in 1960 to 1.5 % in 2000 (FAO 2009). The question is if the postulated global yield growth rate of 1.7 %/year up to 2050 can be reached. Further, a global yield growth rate is an average based on very different situations with respect to technology and the institutional framework. For instance, in Western countries the potential for increasing yields may be quite limited while, on the other hand, fertilization could lead to substantial increases in yield in many locations in Africa and other parts of the developing world.

The assessed increase in production allocated to increase in arable land is, in total, 120 million ha net in the developing world (mainly sub-Saharan Africa and Latin America). If all the additional land were to come from forested areas, that would mean a deforestation rate of 3.1 million ha/year just from expanding agriculture (the current total net deforestation is some 5–6 million/ha/year). For the developed countries, the assessment is that there will be a net decline of 48 million ha in croplands up to 2050 due to the transformation of agricultural land for other land uses (Bruinsma 2009).

The FAO assessment of the potential to increase production is based on the assumption that the area able to be irrigated can expand substantially during the period to 2050. At the same time, organizations like the International Food Policy Research Institute (IFPRI) warn that regions like China (already overusing its water resources) and large parts of Asia should not rely on their current irrigation capacities because of climate change (Nelson et al. 2010). The FAO has not taken climate change into account in its long-term assessment but stresses that it could represent a major risk for long-term food security (as discussed in the coming sections). It is stated in the FAO assessment that climate change could reduce agricultural output in Africa by 15–30 % in the twenty-first century. Moreover, the FAO underlines that investment, productivity increases, area expansions, etc., are not sufficient to harness future food security unless the poverty issue is resolved.

The other threat identified to long-term food security is the use of food crops and cropland for bioenergy production. As outlined in the previous section, the demand for bioenergy is expected to increase substantially. The OPEC (Organization of the Oil Exporting Countries) Fund for International Development (OFID 2009) of the (OPEC) was one of the first to present an integrated study of the impacts of increased use of agriculture feed stock for biofuels on the food sector. In that study, the Global Agro-ecological Zones (GAEZ) model framework was used for the scenario analysis with a time horizon of 2050. The study showed that the biofuel targets of first-generation biofuels discussed worldwide for 2020 would require a cereal use of 150–240 million tons, with huge effects on food prices and potentially an additional 140 million people being put at risk of hunger. The global cultivated land area would have to be increased by 20–40 % in 2020 compared to the baseline scenario if biofuel production were to take place; and this would mean increased deforestation. Despite the fact that OFID (2009) was one of the first approaches to analyzing the field of fuel and food competition globally, there needs to be careful consideration of some of the study's specific shortcomings when interpreting its outcome, such as old and aggregated input data (i.e., climate change data, land cover data), methodological matters (non-peer-reviewed components of the modeling framework), and the domination of first-generation biofuel scenarios.

There are later assessments and scenarios that have been developed with respect to global future food demand. Tilman et al. (2011) have presented new food demand and supply scenarios for 2050 that highlight potential trade-offs between food security, energy production, and climate mitigation. Their first scenario follows a business-as-usual trend, with poorer, lower-yielding nations increasing food production by land clearing and low productivity increases, and richer high-yielding countries increasing their yield through intensification and yield improvements. The second scenario builds on substantial global investments and efficient policy implementations in agricultural technologies, adaptation, and transfers. In both scenarios from Tilman et al., food demand is substantially higher compared to the FAO assessment. In the first case (business-as-usual) demand is met, but only with severe impacts in the form of 1,000 million ha of additional land cleared and global agriculture GHG emissions (in CO_2

equivalents) of 3 Gt/year. In the second scenario (large technology investments) the additional land clearing would be 200 million ha and the GHG emissions 1 Gt/year. However, these scenarios have not assessed neither the size of necessary investments nor where the capital would come from, nor have they identified the policies needed.

The following overall conclusions can be made based on the scenarios investigated:

- *Global agricultural production needs to increase by 70–100 % to feed the population by 2050*
- *Implementation of relevant and effective agricultural policies will be decisive if higher production is to be reached*
- *Higher production might require some 200 million ha of additional arable land by 2050*
- *If grain were to become the major raw material for biofuel production (first-generation ethanol) there could be a need for an additional 300–400 million ha of land on top of the 200 million ha needed for increased food production*
- *The clearing of land (deforestation) just for increased food production could amount to 125–150 million ha by 2050 under particularly extreme assumptions*

4.3 Future Demand for, and Supply of, Forest Industrial Products

In the past, the FAO used production, trade, and prices of the global wood supply to present global studies on long-term consumption of forest products. However, the last such assessment was published in 1998. A number of academic institutions have developed models and scenarios based on official statistics of the long-term consumption, production, and trade of forest products over the years (see Sedjo and Lyon 1996; Hagler 1998; Sohngen et al. 2001; Buongiorno et al. 2003; Kallio et al. 2003; Northway and Bull 2007; Havlik et al. 2011). However, as many of these models lack recent scenarios and outlooks and address specific questions rather than the total demand for, and supply of, forest products, it is difficult to discuss global future demands and supplies based on available scientific studies. Providing alternative sources of data are a number of big consultancy firms that also develop such assessments on a commercial basis. The advantage of these assessments is that the consultancy firms operate close to the markets and can point out factors not visible in official statistics. Thus, in order to base this discussion on the most recent and relevant scenario, Pöyry's assessment of forest product demand for the period 2010–2030 is used (Wintzell 2011).

According to Pöyry's assessment (Wintzell 2011), both the global softwood and hardwood sawnwood markets are expected to grow at a compound annual growth rate (CAGR) of 1 % by 2030, while demand for wood-based panels will have a faster growth of 2.5 %. The growth in demand is mainly projected for Asia-Pacific.

Global demand for paper and paperboard is assessed to have a total annual growth of 1.7 %, with (Japan, North America, and Europe accounting for −0.6 %

and emerging markets for +3.2 %). Nearly all growth takes place in China, Rest of Asia, and Latin America. The softwood kraft pulp, mechanical, and semi-mechanical pulps are assessed to have flat demand development until 2030. A growth is expected in the demand for bleached hardwood kraft, but the big increase in raw material for the paper and paperboard industry is assessed to take place in recovered paper.

All in all, Pöyry foresees an overall growth in demand for forest products that will generate a substantially increased demand for wood by 2030. Compared to 2009, it estimates an increase in industrial wood consumption of nearly 50 %. This corresponds to 700 million m^3, which equates to some 25 million ha of additional fast-growing plantations in the South or 140 million ha in areas with a productivity similar to Sweden's.

Nilsson (2012) has demonstrated that within the same time horizon the supply situation of economically available industrial wood will be extremely tight and decreasing for different reasons in most regions of the world except Latin America and Oceania.

The Pöyry study (Wintzell 2011) suggests that the development of demand for conventional forest products will be flat or will decline in mature markets. It will be very difficult for the forest industry to achieve satisfactory financial results under no-growth markets (Rennel 2010). Faced with market pressures and low profitability, the forest industry in markets that are now mature shows an increased interest in developing new products. The conventional forest industry is only using a fraction of the development possibilities of wood and its fibers and chemicals. A huge amount of different raw materials (platforms) and products can be developed from wood and with a potential for higher value-added. The development of these products is in its infancy (except for advanced biofuels), but there are huge potentials and high expected CAGR (15–20 %). This process can be stimulated by governmental initiatives on the green economy (Nilsson 2011), using policies and instruments similar to those successfully implemented for a transformation from fossil energy to bioenergy. It is too early to make quantitative assessments of the potentials. However, new green production will also use wood and biomass as a raw material. While some new products may replace current ways of using wood, others will use new wood and biomass, thus adding to the competition for biomass and land.

The overall conclusions are summarized in the following:

- *There will be a substantial increase in the global consumption of industrial forest products up to 2030*
- *The estimated increase is projected to be 50 % for industrial wood or 700 million* m^3 *per year by 2030*
- *Nearly all this growth in demand will take place in emerging economies, while in the Western economies demand will be stagnant or negative*
- *The increased demand for biomass for industrial purposes should be seen in relation to the expected increase in bioenergy consumption (see first section on future energy demand)*
- *Global assessments of future industrial wood supply project tight supply conditions*

4.4 Climate Change

Climate change and land use are related to each other in a number of ways. Climate change will affect agriculture across the world through droughts, flooding, declining harvests, loss of productive land, etc. Climate change is also very much related to forest production, deforestation, the use of biomass for energy, etc. The mechanisms of all these changes are both direct, as an impact of actual climate change, and indirect, as consequences of human agency and policy responding to climate change.

Climate scenarios, which may be seen as tools to communicate what is known and less known about the future climate, are associated with many uncertainties. There are many feedback loops in the climate system that are not that well understood, and each climate model has its own built-in, model-specific uncertainty. The most often reviewed scientific climate scenarios are generated by the Intergovernmental Panel on Climate Change (IPCC). The latest assessment, based on a wide range of emission scenarios and global climate models, is the Fourth Assessment Report (IPCC 2007).

Those scenarios predict a global average surface warming of 1.1–6.4 °C, with a best estimate of 1.8–4.0 °C in 2100 (IPCC 2007). Even if the emissions were stabilized today at year 2000 levels, the temperature would still increase by 0.9 °C in 2100. In general, the winters will warm more than the summers. In addition to rising surface temperatures, the scenarios suggest that precipitation will increase at high altitudes and most likely decrease in Central America, Southern Europe, North Africa, Middle East, Southeast Asia, Brazil, South Africa, and Australia. The snow cover and the permafrost depth will decrease, and it is highly likely that semi-arid areas will face decreasing water resources and more droughts. The sea ice of both the Arctic and Antarctic and the Greenland ice sheet will continue to shrink and contribute to sea level rise. More frequent hot extremes, heat waves, heavy precipitation, and more intense tropical cyclones are very likely.

The Copenhagen Accord of 2009 (UNFCCC 2009) includes a commitment to take action to hold the increase in global temperature below 2 °C in order to avoid the *most dangerous impacts of climate change*. However, in the last decades unprecedented increases in GHG emissions have been experienced in both industrialized countries (Kyoto Protocol Annex 1 countries) *and* major emerging economies (non-Annex 1 countries). New science (Smith et al. 2009; Mann 2009) suggests that a temperature increase of 2 °C represents a threshold not between acceptable and dangerous impacts but rather between *dangerous and extremely dangerous* consequences (Anderson and Bows 2010). Moreover, Anderson and Bows (2010) demonstrate that there are small to no possibilities of keeping the rise in global mean surface temperature below 2 °C (but perhaps below 3 °C) due to increased emissions and constraints on mitigation efforts. This means that the impacts of climate change are likely to be more severe than outlined in the IPCC Fourth Assessment Report (2007). A subsequent IPCC (2011) Special Report which focuses on the interaction between weather, extreme climate events (caused by climate change) and vulnerability, warns of more climate-related disasters caused, for example, by storms, floods, droughts, and heat waves.

Following the IPCC's assessments, a number of studies have been presented with respect to climate change and agriculture and food. The International Assessment of Agriculture Knowledge, Science and Technology for Development (IAASTD 2008), initiated by the United Nations Environment Programme, discusses climate change and agriculture; it concludes that climate change is taking place at a time of increasing demand for food, feed, fiber, and fuel and that it has the potential to irreversibly damage the natural resource base on which agriculture depends. The IAASTD assessment states that agriculture contributes in several major ways to climate change and that climate change has adverse impacts on agricultural production. It further foresees increasing water scarcity and changed timing of water availability, which would constrain agriculture production; it also stresses increased extreme events due to climate change that may have significant consequences in all regions for food security and forestry production. The assessment also points out that animal and plant diseases are likely to increase with climate change.

IFPRI has published a study on food security, agriculture, and climate change to 2050 built on scenario analysis (Nelson et al. 2010). The study shows that yields of wheat, maize, and rice are expected to decline substantially due to climate change. Wheat production is, for example, assessed to decline by 1.3–9.0 % in 2030, by 4.2–12.0 % in 2050, and by 14.3–29.0 % in 2100 compared to the year 2000. Some countries are expected to experience a substantial crop area decline during the period 2010–2050 due to a combination of general socioeconomic developments and climate change impacts. Nearly 20 countries are expected to face a decline of more than 1 million ha/country in crop area during this period. China, Russia, and India are all projected to lose land in the 5–20 million ha range. All the countries expected to gain cropland are developing countries (nearly 30), with Brazil and Nigeria gaining 10–14 million ha each. The majority of the gaining countries are in Africa, implying increased deforestation with expansion of agriculture.

Climate change affects not only land availability but also food prices. Several studies show that food prices are likely to increase substantially from today to 2020 and 2050 due to general socioeconomic developments and increased competition for land (OECD and FAO 2011; Hertel et al. 2010). The IFPRI study (Nelson et al. 2010) also takes climate change into account and assesses that food prices will increase by 45–100 % between 2010 and 2050. Of these price increases, 25–65 % will be due to climate change impacts. Food security measured as available average kcal intake per day can, under an optimistic scenario (without climate change), be secured both in developed and developing countries; however, in a pessimistic scenario (with climate change) there will be a decline in available kcal intake between 2010 and 2050 both in developed and developing countries, with resulting increased malnourishment and starvation. IFPRI (Nelson et al. 2010) concludes that during the period 2010–2050, expected income growth, international trade, and targeted agriculture productivity investments can help mitigate the impacts of climate change. Another conclusion is that the climate change impacts up to 2050 are manageable from a food security point of view, but big challenges remain with respect to impacts during the 2050–2100 period. To avoid a disastrous situation during this latter period, actions against climate change have to be implemented now. Other studies (UK Government Office for Science 2011; Willenbockel 2011) show similar results with respect to price developments. However, the UK Government Office for Science (2011) draws a different conclusion as to food supply in 2050. The report concludes:

> future prices will have significant impacts on hunger…climate change makes it harder to address hunger, as it reduces productivity, especially in low-income countries. The food system modeling suggests that within the range of possible scenarios describing the food system over the next 40 years hunger may either markedly decrease or increase. Policy decisions influencing the driving forces of the food system will have very significant effects on hunger and the amount of food produced.

The overall conclusions from the climate change scenarios are presented in the following:

- *With high probability, the world has already locked itself into a situation that will lead to an increase of the average global surface temperature of more than 2 °C in 2100*
- *The climate impacts will probably be stronger than outlined by the IPCC in the 4th Assessment Report* (IPCC 2007)
- *The climate impacts will threaten long-term food security*
- *Because of climate change impacts food prices will rise substantially and increase world hunger*
- *From a land use perspective the most important climate change impact is on agricultural production in low latitude countries in the form of decreased productivity, lost crops due to drought and flooding, and increased pest and diseases, which, in turn, will lead to clearing of land*
- *Policy decisions concerning the future global food system will have decisive effects on future hunger and the amount of food produced*
- *Expected income growth, international trade, and targeted agriculture productivity investments can help mitigate the impacts of climate change*

4.5 Future Land Use Implications

Land use is the crucial link between human and economic activities and nature. All trends discussed so far have the potential to lead to extensive land use changes. These changes can be direct land use changes, yet also have rebound, cascading, or remittance effects. So far the trends have been discussed one by one, in spite of their interrelated nature. The most advanced land use studies aim to integrate several physical, economic, and social parameters and to assess their integrated impacts, including indirect land use changes.

One unwanted land use effect is deforestation. Deforestation contributes, among other things, to climate change by causing 10–15 % of global GHG emissions (e.g., see Van der Werff et al. 2009), changed weather, loss of biodiversity, as well as changed living conditions for and displacement of populations. Because of its centrality to the global land use debate, this issue will be briefly explored before the integrated impacts of those trends are discussed further.

There is no consensus on the magnitude and rate of global deforestation. Several studies explore this question and come to different conclusions (see FAO 2006, 2010; FAO and JRC 2011). These studies suggest that the gross deforestation rate is in the range of 13–15 million ha/year and does not really decline over time. They also indicate that the net deforestation rate is associated with big uncertainties and that no strong trends can be identified. Hansen et al. (2010) found gross forest losses of the magnitude of 20 million ha/year during the period 2000–2005. To understand the forest change dynamics, they argue, the proximate causes of forest cover loss, and the recovery rate of earlier deforested areas have to be taken into account. However, no consensus on causes and impacts of various activities exists. Subsistence farming, logging, fuelwood removal, large-scale commercial agriculture and pasture, as well as charcoal production are mentioned as drivers, but with shifting priority in different studies (UNFCCC 2007; UCS 2011; IFAD 2010; FAO 2010). De Fries et al. (2010) demonstrate that there is a shift in the drivers of deforestation over time. They conclude that urban population growth and agricultural trade now have significant impacts on the deforestation rate. The Union of Concerned Scientists (2011) also stresses that biofuel plantations and palm oil plantations are having, and will have, substantial impacts. Taken together, ongoing deforestation and the expected increase in demand for food and biofuels (based on agricultural crops) suggest a requirement for uncompromising and globally coordinated action to halt deforestation in a foreseeable future. Although most deforestation takes place in semi-tropical and tropical forests, there are "system spillovers" that may influence the conditions for boreal and Nordic forest management. The supply of tropical forests and timber may, for instance, influence the demand for and prices of boreal forest products. Moreover, efforts to mitigate climate change in one part of the world may lead to deforestation somewhere else. Consequently, many integrated land use analyses take deforestation into account.

Since climate change, energy transition, changes in food-, energy-, and forest-production systems are complex processes, their expected accumulated effects are marked by huge uncertainties. The tools to study future land use change are limited by lack of data, little-understood feedback loops, complexity, institutional and societal understanding and responses, etc. Nonetheless, a number of integrated land use analyses have been carried out. A first quantitative global analysis of land use in 2030, integrating forests, agriculture, livestock, and protected areas, was presented by Nilsson and Fischer (2007). The physical global land reserve (i.e., spare productive land suitable for agriculture but not yet cultivated) was assessed at around 700 million ha. An economic and risk dimension was added to the analysis which reduced the accessible land reserve to 295 million ha. By taking into account projected demands for bioenergy, biofuels, food, and industrial forest products, the land balance for 2030 showed a deficit of 250–300 million ha of productive land.

Lambin and Meyfroidt (2011) carried out a major study by compiling and harmonizing all land use studies carried out in recent years to produce global land balance scenarios for 2030. The starting point in the analysis was to assess the extent of unused productive land in the year 2000. This assessment resulted in a high estimate of 445 million ha. The low estimate of global unused productive land is

356 million ha (with added constraints for institutions, land tenure, property and economic rights, political conflicts, and some biophysical factors).

In the next step the additional land needed in 2030 was assessed by taking into account the demand for new agricultural croplands, biofuel croplands, grassing land, urban expansion, industrial forestry expansion, expansion of protected areas, and land lost due to land degradation. Again, a low and high estimate was made. The low estimate is 285 million ha and the high estimate is 792 million ha for 2030. It should be pointed out that these assessments do not take into account climate change impacts, increase in bioenergy use (other than biofuels), feedstock for other bio-products, and socioeconomic dynamics. The resulting land balance for 2030 shows that there will be a land deficit only under the high land demand scenario. In this case there is a deficit of some 350–435 million ha. Thus, in the low land demand scenario, there is sufficient land area available.

However, the authors also did global land base scenarios taking into account what they call unavoidable deforestation. They operated with an accumulated deforestation for the period of 2000–2030 of 152 million ha in a low estimate and 303 million ha in a high estimate. By allowing deforestation, the land deficits will be less than in the alternative no deforestation scenario. With the low estimate of additional land needed and low deforestation rate, there will be no deficits of land at all. Under the assumptions of a high demand for additional land, the deficit will be between 44 and 284 million ha in 2030.

The most advanced integrated land use analysis to date has been carried out by the International Institute for Applied Systems Analysis (IIASA). The Institute has developed a grid-based model system for the analysis of land use changes. The most recent application is work carried out together with the World Wide Fund (WWF) for Nature on future forest development (WWF 2011). The models used by IIASA currently take into account agriculture croplands, biofuel croplands, other wood bioenergy, industrial forest expansion, urban expansion, expansion of protected areas, and land degradation. Climate change, grazing, and socioeconomic developments are under implementation in the model system.

The objective of the WWF (2011) study is to try to find pathways for avoiding deforestation by 2020 under increasing pressure by a growing population, increased food demand, and higher bioenergy and biofuel demands. The scenario analyses are carried out for the period 2010–2050. Four scenarios are compared with a business-as-usual scenario as outlined in Table 4.1.

The WWF business-as-usual scenario generates an accumulated deforestation of 232 million ha and 242 million ha of natural forests converted to managed forests by 2050. The WWF bioenergy scenario shows similar effects on deforestation and a higher degree of forest conversion, whereas the other scenarios generate substantially less deforestation. In terms of GHG emissions, there seem to be considerable positive effects on GHG emission reductions from avoided deforestation in these scenarios (see also Kraxner and Nordström, Chap. 5, this book, and Kraxner et al. 2013). In the WWF scenario package, all the different demands can be met without any land deficits but with a varying degree of deforestation and conversion of natural forests to managed forests. This is because the analytical

Table 4.1 Four scenarios are compared with a business-as-usual scenario to try to find pathways for avoiding deforestation by 2020 under increasing pressure from population growth, increased food demand, and increased bioenergy and biofuel demands (WWF 2011)

Scenario conditions	Accumulated deforestation 2050 (million hectares)	Natural forests converted to managed forests 2050 (million hectares)
Fully renewable energy systems; increased bioenergy feedstock production	245	304
Dietary shift; reduced intake of animal calories	110	241
Far-reaching protection of natural ecosystems	13	265
Zero net deforestation and forest degradation by 2020	56 (up to 2020)	242
Business as usual; no additional policy intervention	232	242

tools used do not allow any "deficits," but highlight trade-offs between the different land use strategies. Further, the WWF scenarios may be combined with each other in different ways so that effects on deforestation and conversion to managed forest will not only accumulate but may also create synergies.

Northway and Bull (2011) also demonstrate huge land use changes up to 2050. Massive transformations of forests to cropland will take place in emerging economies. Roberts (2010) demonstrates strong price increases during the 2000s for land in Brazil, Uruguay, the southern USA, and Malaysia, which indicates that land has already become a scarcer resource. If the Lambin and Meyfroidt (2011) land balance scenarios for 2050 (discussed earlier in this chapter) are adjusted to take climate- and energy-related demands for additional crop- and forestlands into account, the deficit of land will be 128–1,048 million ha by 2050, if no deforestation is allowed. If deforestation is allowed, the deficit of land in the global balance will be up to 528 million ha, given the high estimate of demand for land. Consequently, there is a high risk of land use conflicts by 2050 under all the scenario conditions used in this application. The conclusion of these exercises is that *the accumulated effects of these global trends will cause land scarcity and possible land conflicts.*

However, this is not the complete picture. There are huge uncertainties involved in the basic data. The climate assessments, the calculations of land used/needed for agriculture, the assessments of deforestation rates and the remote sensing data are all associated with big uncertainties (Fritz et al. 2011; Kraxner 2011; UK Government Office for Science 2011; White et al. 2011). Given the magnitude of these uncertainties, a possible conclusion is that almost nothing definitive can be stated about future land use and land use conflicts based on the scenarios presented in this chapter. Lambin and Meyfroidt (2011) chose to conclude that the land issue can *be harnessed if land use is understood as being part of open and complex human/nature/environmental systems.* If the right policies and instruments are implemented, there will be no land use problems by 2050. IIASA and WWF (WWF 2011) argue the same way: *with better governance the world would have enough productive forest land and agriculture land available for the demand for food, wood products, and bioenergy in 2050.* Consequently, appropriate and efficient policy is key (see also Shell 2008; Nelson et al. 2010; the UK Government Office for Science 2011).

Obviously, it is possible to come to a particularly optimistic conclusion; the land use issues can be handled up until 2050 without major conflicts as long as the right policies and measures are implemented. So, with this interpretation, *we can do it;* we can avoid land use conflicts by 2050. *Will we do it?* Based on earlier experiences of governance failing to solve complex problems, a reasonable guess is that *we will not.*

Global society has not been very efficient in implementing the policies required to mitigate climate change. Anderson and Bows (2010) and IEA (2011) demonstrate that there was a window for halting the average surface warming below 2 °C, but policymakers did not manage to implement the required policies and now the window is possibly closed due to mitigation constraints. Shell (2011) concludes that society is currently moving on a trajectory which is closer to business as usual, rather than addressing the right issues with respect to the needed transformation of the energy system. The Global Energy Assessment (2012) proposes that the higher future energy consumption is (and global society is on an increasing trajectory) then the more difficult it will be to find sustainable transformations of the energy system. It has been known since the 1980s that the growth rates in agricultural yield for major crops have been declining due to failed policies and that the trajectory is still declining. Deforestation has been a known problem since the 1970s, and the recent assessment indicates that the gross deforestation rate is still not decreasing.

Each of the global trends discussed here requires difficult, complex, efficient, and society-wide policies and measures that will be more difficult to implement as time goes on. Therefore, a third and different interpretation can be made from the scenario exercises. Despite policies and measures to implement these measures being known, global society will not be able to do it in time because of political and societal conflicts, institutional constraints, property and economic rights, etc. Lack of steering, in combination with the global trends themselves, will generate problems of immense complexity, with cascading sets of interactions, trade-offs, and synergies that will be difficult (if not impossible) for decision makers to solve (Havlik et al. 2011; Obersteiner 2011).

The overall conclusions of the integrated land use analyses are:

- *There are huge uncertainties in the data on which the scenarios are based.*
- *Timely implementation of appropriate policies is key to avoiding problems of immense complexity that may be difficult, or impossible, to solve.*
- *Based on previous experience of governance failures, there is a high risk that the steps needed will not be taken in time.*
- *Therefore, it is likely that global society will face a situation in line with the high land demand estimates in the scenarios: a deficit of productive land in the range of 500–1000 million ha and severe land-use conflicts in 2050.*

4.6 Conclusions in a Nordic and Boreal Context

It can now be concluded that a number of interrelated global trends will drive global land use change, some of which will have direct effects on the conditions for future Nordic forest use. The accumulated effects of other trends on global land use will influence the Nordic situation. Whatever happens, the Nordic forest sector must relate to these developments and position itself in a changing global environment. From a Nordic perspective, the most important conclusions of the scenario exercise, as shown in earlier sections of this chapter, highlight a growing demand for forest resources, bioenergy, and land, a changing demand for forest products, and a strong need for policy that tackles the challenges of climate change in an integrated manner. So, how can the Nordic states and forest sector contribute, and relate to, the policy development needed? What does growing demand for biomass and global land scarcity mean to Nordic forest management, and what opportunities and challenges do the changing forest products markets offer the Nordic forest sector?

One clear conclusion is the need to develop efficient policies and measures. The difference between disaster and sustainability in 2050 resides in the right policies being implemented speedily and efficiently. It means that the Nordic states and forest sector actors have to establish policies that take institutional, societal, and political constraints into account. The states must develop policy to mitigate climate change and transform energy systems, while addressing the role of forests within these processes. Forest sector actors may choose to take a proactive or passive role. Improved governance of the Nordic forest sector will be critical to their capacity to carry through this transition process in a planned and organized way. An important dimension of the new governance and planning situation is the assessments of possible trade-offs between the production of products and delivery of other services by the land resource in the future. These trade-offs often cause institutional, societal, and political conflicts (see Havlik et al. 2011; and Chaps. 6, 8, and 10, this book). As forests are expected to deliver a broader range of products and services in the future, interest groups outside of the forest sector will intervene more and conventional forest sector governance and planning will probably become obsolete. The concept needed will be multisectoral and integrated governance and planning of land resources.

The scenario assessments discussed earlier show convincingly that global demand for forest resources and land is expected to increase. There is a high probability that productive and unproductive land will become a scarce resource in the future. The most dramatic land use changes and the dominating land crunch will take place in the South. However, in order to position itself in relation to a changing global land use situation and establish viable strategies, the Nordic forest sector must understand forest sector development in the South and in the BRIC countries. More integrated research is needed to clarify the relationships between, on the one hand, global policy and land use, and, on the other hand, Nordic forest management and governance. Increasing global demand for forest resources may, for example, make the boreal zone more attractive for investments related to climate change, energy, environmental, and population change. The multiple demands directed toward biomass also signal increasing real prices for forest products and increasing economic and social contributions to society from the forest sector. Increased real prices provide an opportunity to realize new governance and new policies for the sector. Moreover, in order to better account for societal impacts and benefits on a landscape level, agent-based modeling as an important part of integrated assessment might need to be applied.

The scenarios demonstrate further that the global forest sector is in the process of transformation. They also indicate that a transition/transformation of the Nordic forest sector will not be easy. The conventional markets for both standard products and volumes produced in the boreal region are contracting over time and it will be difficult for them to be competitive with these products on other markets. Therefore, the boreal forest industry must change its business concept to new products with higher value-added and produce more value from less. The forest industry in the Nordic region is currently utilizing only a fraction of the wood and fiber potential. New products could be biorefinery products, green chemicals, advanced fuels, advanced constructions, smart packaging, and advanced hygienic and health products. These new products often have a value-added that is 10–100 times higher than for conventional forest products. Another chapter by Nilsson (Chap. 9) in this book explores how Canada, a major forest- producing nation, is facing similar challenges because of delayed reaction toward the new challenges including, inter alia, a shift from classical forest products, the financial crisis, and environmental issues such as climate change.

Many observers argue that the challenges to the over-consumption of resources are so big that sustainable development requires a fundamental transformation to a green economy (e.g., UNEP 2011a, b). The natural resource-based part of the economy, not the least the forest sector, is well positioned to take the lead. Nilsson (2011) points out that this is of vital interest for the economy in general. A change in this direction would help to transform the conventional forest sector. The institutions of the forest sector could take the initiative to demonstrate their possible contributions to a green economy and drive the policymakers in this direction. The Nordic regions and their forest sectors have already entered this transformation, as evidenced by, for example, their strongly increased use of bioenergy and renewable energies, especially from the forest sector. This role has to be further strengthened by moving from low value-added bioenergy (like chips and pellets) and into advanced bioenergy with higher value-added and energy efficiency.

References

Alexandratos, N. (2009). World food and agriculture to 2030/2050: Highlights and views from mid-2009. In *Proceedings, FAO expert meeting, how to feed the world in 2050*. Rome: FAO.

Anderson, K., & Bows, A. (2010). Beyond "dangerous" climate change: Emission scenarios for a new world. *Philosophical Transactions of the Royal Society, A2011*(369), 20–44.

Bruinsma, J. (2009). *The resource outlook to 2050. By how much do land, water and yields need to increase by 2050?* Paper presented at FAO expert meeting, 24–26 June 2009, How to feed the world in 2050, FAO, Rome, Italy.

Buongiorno, J., et al. (2003). *The global forest products model: Structure, estimation, and applications*. San Diego/Burlington: Academic.
DeFries, R., Rudel, T., Uriarte, M., & Hansen, M. (2010). Deforestation driven by urban population growth and agriculture trade in the twenty-first century. *Nature Geoscience, 3*(3), 178–181.
Ecofys and WWF. (2010). *The energy report. 100% renewable energy by 2050*. Gland: Ecofys and WWF International.
FAO. (2006). *Global forest resource assessment 2005* (FAO forestry paper 147). Rome: FAO.
FAO. (2009). *How to feed the world in 2050*. Rome: FAO.
FAO. (2010). *Global forest resource assessment 2010* (FAO forestry paper 163). Rome: FAO.
FAO and JRC. (2011). *Global forest land-use change from 1990 to 2005. Initial results from a global remote sensing survey*. Rome: FAO.
Fritz, S., See, L., McCallum, I., Schill, C., Obersteiner, M., van der Velde, M., Böttcher, H., Havlik, P., & Achard, F. (2011). Highlighting continued uncertainty in global land cover maps for the user community. *Environmental Research Letters, 6*(4), 044005.
Global Energy Assessment. (2012). Cambridge: Cambridge University Press.
Greenpeace. (2010). *Energy revolution. A sustainable world energy outlook*. Amsterdam: Greenpeace International.
Hagler, R. (1998). *The global timber supply/demand balance to 2030: Have the equations changed?* A multi-client study by Wood Resources International, Reston.
Hansen, M. C., Stehman, S. V., & Potapov, P. V. (2010). Quantification of global gross forest cover loss. Early edition of *Proceedings of National Academy of Sciences*, USA. www.pnas.org/cgi/doi/10.1073/pnas.0912668107
Havlik, P., Schneider, U. A., Schmid, E., Böttcher, H., Fritz, S., Skalský, R., Aoki, K., De Cara, S., Kindermann, G., Kraxner, F., Leduc, S., McCallum, I., Mosnier, A., Sauer, T., & Obersteiner, M. (2011). Global land-use implications of first and second generation biofuel targets. *Energy Policy, 39*, 5690–5702.
Hertel, T. W., Tyner, W. E., & Birur, D. K. (2010). Global impacts of biofuels. *Energy Journal, 31*(1), 75–100.
IAASTD. (2008). Synthesis report of the International Assessment of Agriculture Knowledge, Science and Technology for Development.
IEA. (2011). *World energy outlook 2011*. Paris: International Energy Agency.
IFAD. (2010). *Global poverty report 2010*. Rome: International Fund for Agricultural Development.
IPCC. (2007). *Climate change 2007: Synthesis report*. Cambridge: Cambridge University Press.
IPCC. (2011). *The IPCC special report on managing the risks of extreme events and disasters to advance climate change adaptation*. Geneva: IPCC.
Kallio, M. A. I., Miseyev, A., & Sohlberg, B. (2003). *The global forest sector model EFI-GTM. Internal report 15*. Joensuu: European Forest Institute.
Kraxner, F. (2011, September 14). *Global shifts of land use: Natural resources and the future*. Presentation at future agriculture, Swedish University of Agricultural Sciences, Uppsala, Sweden.
Kraxner, F., Nordström, E.-M., Havlík, P., Gusti, M., Mosnier, A., Frank, S., et al. (2013). Global bioenergy scenarios – Future forest development, land-use implications and trade-offs. *Biomass and Bioenergy, 57*, 86–96.
Lambin, E. F., & Meyfroidt, P. (2011). Global land use change, economic globalization, and the looming land scarcity. *Proceedings of the National Academy of Sciences, 108*(9), 3465–3472.
Langeveld, J. W. A., Dixon, J., & Jaworski, J. F. (2010). Development perspectives of the biobased economy: A review. *Crop Science, 50*(Supp. 1), S-142–S-151.
Mann, M. E. (2009). Defining dangerous anthropogenic interference. *Proceedings of the National Academy of Sciences, 106*, 4065–4066.
Nelson, G. C., et al. (2010). *Food security, farming, and climate change to 2050. Scenarios, results, policy options*. Washington, DC: International Food Policy Research Institute.

Nilsson, S. (2011, October 4–7). *Contributions of the forest sector to economic development and transition towards a green economy*. Presentation at the international conference contribution of forests to a green economy, German Government, Bonn, Germany.
Nilsson, S. (2012). Availability of cultivated land to meet the expected demand in food, fiber, and fuel. In: Royal Swedish Academy of Agriculture and Forestry, The global need for food, fibre and fuel: Land use perspectives on constraints and opportunities in meeting future demand. *Tidskrift*, 151 (4), The Swedish Academy of Agriculture and Forestry.
Nilsson, S., & Fischer, G. (2007). *The three Fs: Food, fiber, and fuel*. Presentation at IIASA 35 year anniversary conference on global development: Science and policies for the future, IIASA, Laxenburg, Austria.
Northway, S., & Bull, G. (2007). *Review of the capabilities of the international forest and forest products trade model*. Vancouver: University of British Columbia.
Northway, S., & Bull, G. (2011, February 24). *Global trends in the forest sector: Land use, products and trade*. Presentation at the ABCPF forestry conference "Wood is Good," University of British Columbia, Vancouver, Canada.
Obersteiner, M. (2011, Summer). Green food systems for the 9 billion. *Options*, IIASA, Laxenburg, Austria.
OECD and FAO. (2011). *OECD-FAO agricultural outlook 2011–2020*. Paris: OECD.
OFID. (2009). *Biofuels and food security. Study prepared by IIASA*. Vienna: OFID.
Rennel, J. (2010). *Long-term value creation and value destruction. Successful and failed investments in the forest industry*. Stockholm: Spearhead Production. [in Swedish]
Riahi, K., et al. (2012). Energy pathways for sustainable development (chapter 17). In *Global energy assessment*. Laxenburg: IIASA.
Roberts, D. (2010, September 30–October 1). *The global competition for land: The 4 Fs (Food, Feed, Fiber, and Fuel)*. Presentation at 1st Latin America forest industry conference, held Sao Paulo, Brazil, Cambridge Forest Products Associates, MA, USA.
Sedjo, R. A., & Lyon, K. S. (1996). *Timber supply model 96: A global timber supply model with a pulpwood component* (Discussion paper 96–15). Washington, DC: Resources for the Future.
Shell. (2008). Energy. In *Shell energy scenarios to 2050*. The Hague: Shell International.
Shell. (2011). Signals & signposts. In *Shell energy scenarios to 2050*. The Hague: Shell International.
Smith, J. B., et al. (2009). Assessing dangerous climate change through an update of the Intergovernmental Panel on Climate Change (IPCC) "Reasons for concern". *Proceedings of the National Academy of Sciences, 106*, 4133–4137.
Sohngen, B., Mendelsohn, R., & Sedjo, R. (2001). A global model of climate impacts on timber markets. *Journal of Agriculture Resource Economics, 26*, 326–343.
Tilman, D., Balzer, C., Hill, J., & Befort, B. E. (2011, December 13). Global food demand and the sustainable intensification of agriculture. *Proceedings of the National Academy of Sciences, 108*(50), 20260–20264.
UCS. (2011). *The root of the problem: What's driving tropical deforestation today?* Cambridge, MA: Union of Concerned Scientists.
UK Government Office for Science. (2011). *Foresight. The future of food and farming*. London: The Government Office for Science.
UNEP. (2011a). *Towards a green economy. Pathways to sustainable development and poverty eradication*. Nairobi: UNEP.
UNEP. (2011b). *Forests in a green economy. A synthesis*. Nairobi: UNEP.
UNFCCC. (2007). *Investment and financial flows to address climate change*. Bonn, Germany: http://unfccc.int/files/essential_background/background_publications_htmlpdf/application/pdf/pub_07_financial_flows.pdf
UNFCCC. (2009). *Copenhagen Accord. FCCC/CP/2009/L.7*. Copenhagen: United Nations Climate Change Conference.
Van der Werff, G. R., et al. (2009). CO_2 emissions from forest loss. *Nature Geoscience, 2*(11), 737–738.

WBGU. (2010). *Future bioenergy and sustainable land use* (Flagship Report). London: The German Advisory Council on Global Change (WBGU)/Earthscan.
White, T., et al. (Eds.). (2011). *Greenhouse gas inventories. Dealing with uncertainty*. Dordrecht: Springer.
Whiteman, A. (2010, April 28–29). *The context for forest & energy: Setting the scene. The global potential for energy production from forests*. Paper presented at ICF national conference 2010: "Forests and Energy—Maximizing the Potential," Newcastle, UK.
Willenbockel, D. (2011). *Exploring food price scenarios towards 2030 with a global multi-region model*. Oxfam Research Report.
Wintzell, J. (2011, September 27). *Global demand for wood and fibre in the next 20 years*. Presentation at the seminar "The global need for food, fibre and fuel: Is there enough land to satisfy the demands?", The Royal Swedish Academy of Agriculture and Forestry, Stockholm, Sweden.
WWF. (2011). *Living Forest Report*. Gland: World Wide Fund for Nature (WWF).

Chapter 5
Bioenergy Futures: A Global Outlook on the Implications of Land Use for Forest-Based Feedstock Production

Florian Kraxner and Eva-Maria Nordström

Preservation of biodiversity and reduction of deforestation are considered to be key elements whenever an increased use of bioenergy in the future is addressed. This chapter presents different scenarios to 2050 for global feedstock supply for the production of bioenergy under specified social and environmental safeguard provisions. The scenarios were developed through the application of an integrated global modeling cluster; they show that biomass for bioenergy will, to a large extent, be sourced from the conversion of unmanaged forest into managed forest, from new fast-growing short-rotation plantations, and from intensification of land use. Depending on the underlying scenario, zero net deforestation by 2020 might be reached and upheld with only a minor expansion into managed forests. Results further indicate that with rising populations and projected consumption levels, there may not be enough land to simultaneously conserve natural areas completely, halt forest loss, and switch to 100 % renewable energy, which will make difficult trade-offs necessary. Future food and energy demands would lead to acute land competition and increased pressure on agricultural land and water resources. Managed boreal forests are likely be an important source for bioenergy feedstock and, especially in the tropical regions, it is important to achieve a controlled conversion from unmanaged to sustainably managed forest as well as increased protection of areas for biodiversity.

F. Kraxner (✉)
Ecosystems Services and Management Program (ESM), International Institute for Applied Systems Analysis (IIASA), Laxenburg, Austria
e-mail: kraxner@iiasa.ac.at

E.-M. Nordström
Department of Forest Resource Management,
Swedish University of Agricultural Sciences (SLU), Umeå, Sweden

Ecosystems Services and Management Program (ESM), International Institute for Applied Systems Analysis (IIASA), Laxenburg, Austria

E. Westholm et al. (eds.), *The Future Use of Nordic Forests*,
DOI 10.1007/978-3-319-14218-0_5

Photo 5.1 Along the road to the power station

5.1 Future Bioenergy Feedstock in a Global Perspective

Future sustainable use of the planet's resources calls for renewable and carbon-neutral energy sources to meet increasing demands for energy. According to the International Energy Agency (IEA), there is a major opportunity to reduce fossil CO_2 emissions in transitioning to alternative sources for energy production, which includes the use of biomass from forests or agricultural crops (IEA 2010). Emissions from biomass are generally accounted for as carbon-neutral, but whether biomass is entirely carbon-neutral is disputed and very much dependent on system boundaries in time and space; in addition to emissions from indirect land use change (iLUC), emissions come from processes like cultivation, transport, or fuel production (Kraxner et al. 2013).

In 2011 more than 60 countries had some type of national renewable energy target or support policy, according to the Global Renewable Policies and Measures Database. Climate change mitigation, energy security, and protection of national industries are the main rationales for supporting renewable energy. However, the extent of social and environmental regulations for the production of bioenergy feedstock varies greatly between countries. Many developed and developing countries have ambitious bioenergy targets but lack sound supporting legislation (Kraxner et al. 2013). Where legislation exists, it is often confused, fails to address socioeconomic and environmental aspects properly, and may create perverse incentives (Jull et al. 2007). Moreover, country-level requirements for GHG emission reductions are highly variable, as is the assessment of compliance. In most countries forestry legislation does not contain specific regulations concerning harvesting and use of forest biomass for bioenergy (Stupak et al. 2007).

Deficiencies in policies, legislation, and management guidelines in both developed and developing countries indicate that, especially at the global level, basic social and environmental values are at serious risk from increasing bioenergy production. Pristine forests, biodiversity, agricultural land, and soil and water resources will all be under additional pressure if there is a substantial increase in the use of biomass from agriculture, forestry and waste for energy production. Such a development may also counteract other environmental policies and objectives, such as waste minimization or ecological farming. Lack of proper planning and management of feedstock production could also have severe socioeconomic effects, such as conversion of farmland and forest at the expense of small farmers and people living in the forest, concentration of land ownership, increasing food prices, and additional pressure on food supply in already vulnerable regions (FAO 2008).

Based on these insights, we conclude that it is of utmost importance to define and analyze scenarios of global feedstock supply for bioenergy production in order to identify boundaries for future development and guide further research and policy-making. Economic development, population growth, and social and environmental safeguard provisions should all be taken into account in this process (Kraxner et al. 2013). However, few studies to date have addressed this issue on a global scale (Heistermann et al. 2006). The reason is that this kind of analysis calls for integrated modeling; that is, an interdisciplinary approach that combines economic and bio-physical land use models. For large-scale and global analysis of land-based sectors, general and partial equilibrium economic models are used. In integrated modeling, a biophysical land use model is usually linked to the equilibrium model to provide information on supply constraints and on the actual, spatially explicit effects of land use change processes. A limited number of equilibrium models of truly global scope have been used for modeling land use and land use change (Lotze-Campen 2008; Heistermann et al. 2006). Most of these are focused on agriculture and only a few include the forest sector (Buongiorno et al. 2003; Hertel 1997; Cardellichio et al. 1989). There have been very few attempts to use integrated modeling for a global and spatially explicit assessment of bioenergy feedstock taking into account both the forestry and agricultural sectors (e.g., Kraxner et al. 2013; Havlik et al. 2011).

In this chapter our aim is to provide an outlook on the potential feedstock for bioenergy with a global perspective using an integrated modeling approach. Scenarios similar to those presented in this chapter were developed jointly with the World Wide Fund for Nature (WWF) and used for analysis of future forest development with focus on deforestation (Kraxner et al. 2013; Taylor 2011a, b). In this chapter we provide additional results and interpretations relevant to an outlook on bioenergy feedstock. The aim of such a global outlook is to frame the boundaries for lower-scale assessments, to justify research on bioenergy at various scales, and to identify potential trade-offs to be considered in future research. In addition, we examine some possible implications for boreal and Swedish forests.

5.2 Modeling Biomass Supply at Global Scale: An Integrated Modeling Approach

5.2.1 The IIASA Model Cluster

The models used in this study, GLOBIOM (Global Biosphere Management Model), EPIC (Environmental Policy Integrated Climate), G4M (Global Forestry Model), and POLES (Prospective Outlook on Long-term Energy Systems), form an integrated modeling framework that has long been used at IIASA for scenario analysis. The economic model GLOBIOM (Havlík et al. 2011) bases its crop and forest sector details on biophysical parameters supplied by the more specialized models: G4M (Kindermann et al. 2008) for forestry and EPIC (Izaurralde et al. 2006; Williams 1995) for agriculture. GLOBIOM is further linked to the global energy model POLES of the European Union Joint Research Centre (JRC; Russ et al. 2007) for regions outside Europe and the PRIMES model (Capros et al. 2010) for EU27 countries through information on macroeconomic indicators and bioenergy demand. For a more detailed description of the modeling framework, see, for example, Havlík et al. (2011) and www.globiom.org.

5.2.2 Scenario Settings

The five scenarios we focus on here are summarized in Table 5.1.

As described in Table 5.1, the High bioenergy scenarios assume 100 % renewable energy by 2050 as envisioned by the Ecofys Energy Model (Singer 2011).

Table 5.1 Overview of the five scenarios

Scenario name	Description
Baseline	"Business as usual" future development is in line with historical trends:
	Land use change takes place to supply a growing global population with food, fiber, and fuel, and is poorly planned
	Population growth and increase in GDP are drawn from the European Commission (EC 2011)
	Agricultural productivity gains and human diet changes are based on FAO data on historical trends (Grethe et al. 2011; FAO 2006)
	No restriction on deforestation
Low bioenergy	Like the *Baseline scenario*, characterized by:
	Bioenergy feedstock demand is fixed at the level of 2010
	No restriction on deforestation
High bioenergy	Like the *Baseline* scenario, characterized by:
	More ambitious GHG emission reduction targets are assumed as the projection of bioenergy demand is based on the "global 2 °C scenario" derived from the POLES model (EC 2011; Russ et al. 2007). This is an approximation of the bioenergy demand by 2050 in the 100 % renewable energy vision by the Ecofys Energy Model (Singer 2011)[a]
	Higher carbon price is assumed (the carbon price is endogenously computed by the POLES model as the marginal cost of restricting aggregate GHG emissions)
	No restriction on deforestation
High bioenergy without deforestation	Like the *Baseline* and *High bioenergy* scenarios, characterized by:
	More ambitious GHG emission reduction targets are assumed, as the projection of bioenergy demand is based on the "global 2 °C scenario" derived from the POLES model (EC 2011; Russ et al. 2007). This is an approximation of the bioenergy demand by 2050 in the 100 % renewable energy vision by the Ecofys Energy Model (Singer 2011)
	Higher carbon price is assumed (the carbon price is endogenously computed by the POLES model as the marginal cost of restricting aggregate GHG emissions)
	With restriction on deforestation through the "zero net deforestation and forest degradation by 2020" (ZNDD) target
Biodiversity without deforestation	Like the *Baseline* scenario, characterized by:
	Large areas of natural ecosystems are protected if identified as important for biodiversity by certain conservation mapping processes (UNEP-WCMC/IUCN WCPA 2010; BirdLife International 2008; Ricketts et al. 2005; Mittermeier et al. 2004; Olson and Dinerstein 2002; Duellman 1999; WWF/IUCN 1994)
	Current land uses (e.g., agriculture or forestry) in these protected areas remain constant
	With restriction on deforestation through the "zero net deforestation and forest degradation by 2020" (ZNDD) target

[a]This scenario projects demand for bioenergy from land-based feedstocks in 2050 of 71.4 EJ, of which 16 EJ are liquid biofuels

We should keep in mind that the High bioenergy scenarios do not assume 100 % bioenergy, as bioenergy is only one element in the total renewable energy picture, which also comprises hydropower, wind power, solar, and geothermal energy. Nevertheless, the level of renewables is high in these scenarios compared with, for example, the Global Energy Assessment scenarios which propose that renewable energies could account for 30–75 % of global primary energy by 2050 (GEA 2012). We did, however, choose to consider a high level of renewable energy production in order to create scenarios covering a wide span of bioenergy production to allow us to analyze a wide range of potential land use implications.

The projections of GHG emissions include emissions from land use within the agricultural, forestry, and bioenergy sectors. Thus, temporary carbon sequestration in wood products with a time lag in emissions is taken into account, as are lifecycle GHG savings from the substitution of biofuels for fossil fuels.

A substantial component of our scenarios is the vision of "zero net deforestation and forest degradation by 2020" (ZNDD) in accordance with WWF's Living Forests Report (Taylor 2011a). "Zero net deforestation and forest degradation by 2020" means that there is no net forest loss through deforestation and no net decline in forest quality through degradation after 2020. Thus, for our scenarios we compare a future development of feedstock under (i) the assumption that there are no restrictions with respect to deforestation, except for protected areas; (ii) the assumption that there is a strong restriction on deforestation. Note that plantations are not equated with natural forests, as many ecosystem services are diminished when a plantation replaces a natural forest.

5.3 What Do the Bioenergy Scenarios Tell Us?

5.3.1 Bioenergy Potential Under Different Scenarios

According to IEA (2012) the total primary energy supply in 2010 was 12,717 Mtoe, and 1,278 Mtoe of this was from biofuels and waste. From 2020 and onwards, bioenergy production will increase more under the *High bioenergy* and *High bioenergy without deforestation* scenarios than under the *Baseline* and *Biodiversity without deforestation* scenarios (Fig. 5.1), mainly due to a higher carbon price and more ambitious emission reduction targets under the *High bioenergy* and *High bioenergy without deforestation* scenarios. Note that about the same total bioenergy production is generated under the *High bioenergy* scenarios, irrespective of whether there is a restriction on deforestation or not. However, if there is no restriction on deforestation, most of the bioenergy will be sourced in unmanaged or pristine forest, while under the "no deforestation" target most of the wood-based bioenergy production will take place in managed forest and short rotation plantations and also partly through a conversion from unmanaged to managed forest.

Bioenergy for heat and power will be increasingly important in the future and will form a large part of the total bioenergy. Figure 5.2 illustrates the production

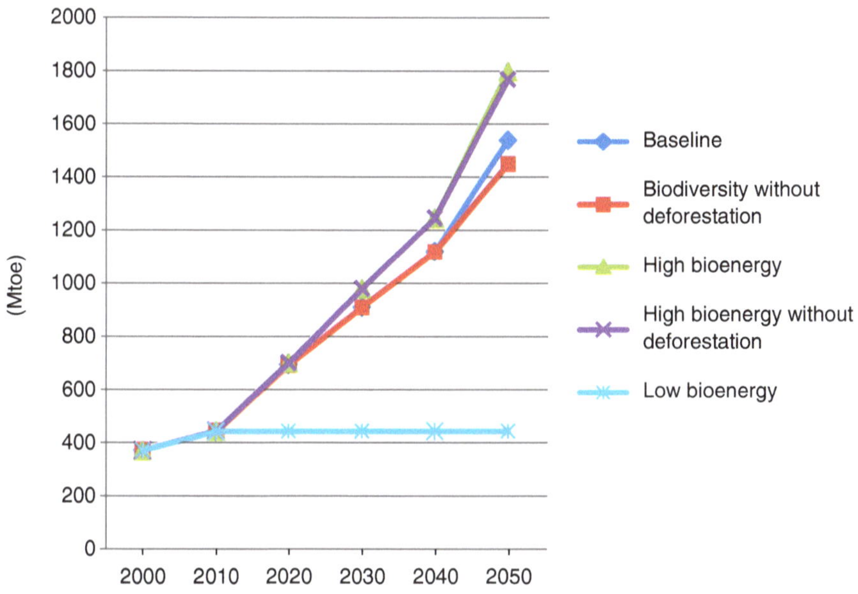

Fig. 5.1 Total production of bioenergy 2000–2050 under the different scenarios (Kraxner et al. 2013)

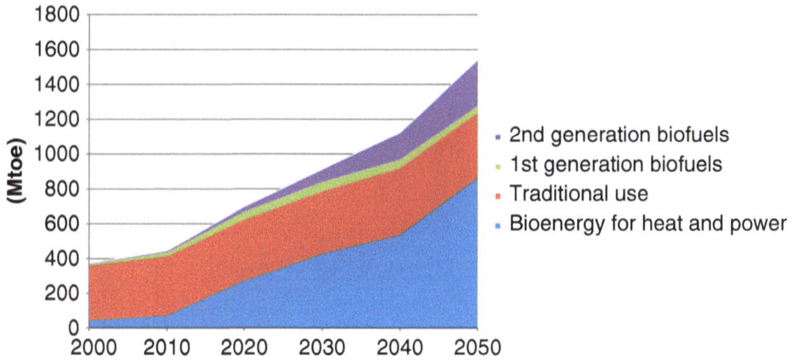

Fig. 5.2 Production share of different types of bioenergy 2000–2050 under the *Baseline* scenario

share of different types of bioenergy under the *Baseline scenario*. All but the *Low bioenergy* scenario follow a similar pattern. Liquid biofuels can be divided into first-generation biofuels, which are produced from crop-based biomass, for example, sugarcane ethanol or rapeseed biodiesel, and second-generation biofuels, which are wood-based.

In 2010 the production of first-generation biofuels was 26 Mtoe. The production will almost double in 2030–2040 in all but the *Low bioenergy* scenario, and then decrease. Around 2030–2040 a shift from first- to second-generation biofuels takes place. The production of second-generation biofuels was only around 4 Mtoe in 2010 but will increase considerably in all scenarios except for *Low bioenergy*. The

Photo 5.2 Chopped for district heating

increase is especially notable under the *High bioenergy* and *High bioenergy without deforestation* scenarios where production rises 150 % between 2040 and 2050. This shift to second-generation biofuels means a shift in feedstock from mainly agricultural crops to forest or plantation-based feedstock, with an overall intensification of agriculture and forestry expected. There will be higher pressure on unmanaged forest to be converted into managed forest to provide bioenergy feedstock. Both natural and cultivated land will also be converted into plantations for the same purpose.

5.3.2 Land Use Change Under Additional Bioenergy Demand

Land use change will take place for many reasons. In this section we focus on land use change due to bioenergy demand. The models used handle six land cover classes. In the base year 2000, the areas of each land cover class were: 906 million hectares (ha) cropland, 719 million ha managed forest, 3,146 million ha unmanaged forest, 47 million ha short rotation coppice, 1,157 million ha grassland and 2,286 million ha other natural vegetation. Figure 5.3 shows the total cumulative land use change under the *Baseline* scenario. Unmanaged forest and other natural vegetation will decrease, and the net forest change will be just below 300 million ha in 2050.

By comparing the land use change under the *Low bioenergy* scenario to the land use change in the other scenarios, the effect of increased bioenergy feedstock production compared to the 2010 level can be assessed. First, in the *Baseline* scenario, we see that the areas of short rotation coppice and managed forest will be relatively higher due to additional bioenergy feedstock production Fig. 5.4. The areas of other

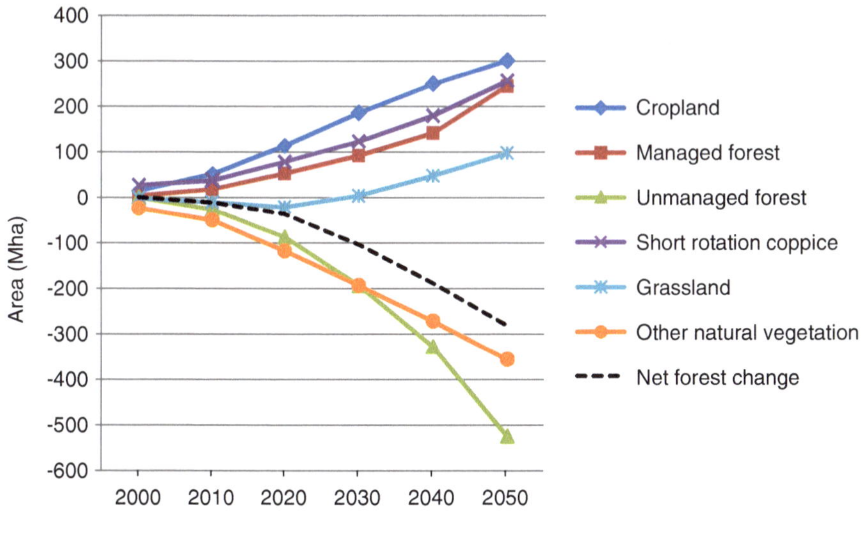

Fig. 5.3 Total cumulative land use change under the *Baseline* scenario (Kraxner et al. 2013)

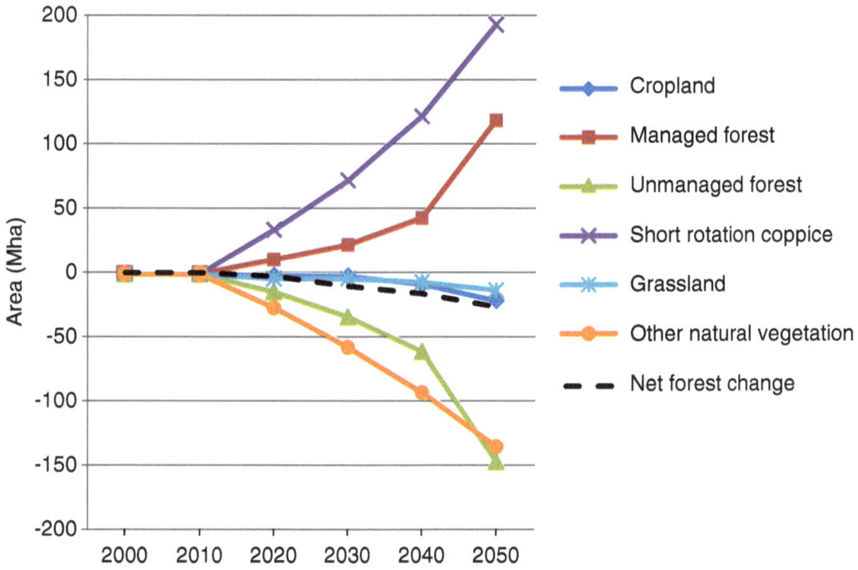

Fig. 5.4 Cumulative land use change and net forest cover change (managed + unmanaged forest area) caused by additional bioenergy production under the *Baseline* scenario (compared to the 2010 level of bioenergy production) (Kraxner et al. 2013)

natural vegetation (i.e., mainly dry shrubland areas) and unmanaged forest will decrease. Cropland and grassland areas will also decrease slightly. Thus, a large part of the total increase in area of short rotation coppice can be explained as being caused by increased future bioenergy demand. Similarly, about half of the total increase in managed forest area is also caused by bioenergy demand. Figure 5.4 also shows that there is a loss of total forest area (net forest change) of 26 million ha which is caused by additional bioenergy production when there is no restriction on deforestation. However, bioenergy is not the direct driving factor behind the loss of pristine forest and other natural habitat. The exploitation of these habitats is caused by other factors, for instance, expansion of agricultural land for food production and inefficient land use. Land use change caused by bioenergy production displays a similar pattern under the *High bioenergy* scenario, although the changes are of a larger magnitude.

A comparison between the *Baseline* scenario and the *Biodiversity without deforestation* scenario shows that the target of avoiding deforestation creates a change among the different land use types (Fig. 5.5). Areas of other natural vegetation, grasslands, and croplands decrease, while areas of short rotation coppice and managed forest increase. An expansion of unmanaged forest also takes place with a certain time lag in formerly managed forests and croplands where management has ceased. Overall, there is an increase in total forest area due to the restriction on deforestation (see net forest change in Fig. 5.5). The protection of biodiversity within pristine and other types of forest would clearly be at the expense of, for example, grassland and savannah (which are mainly located in the southern hemisphere).

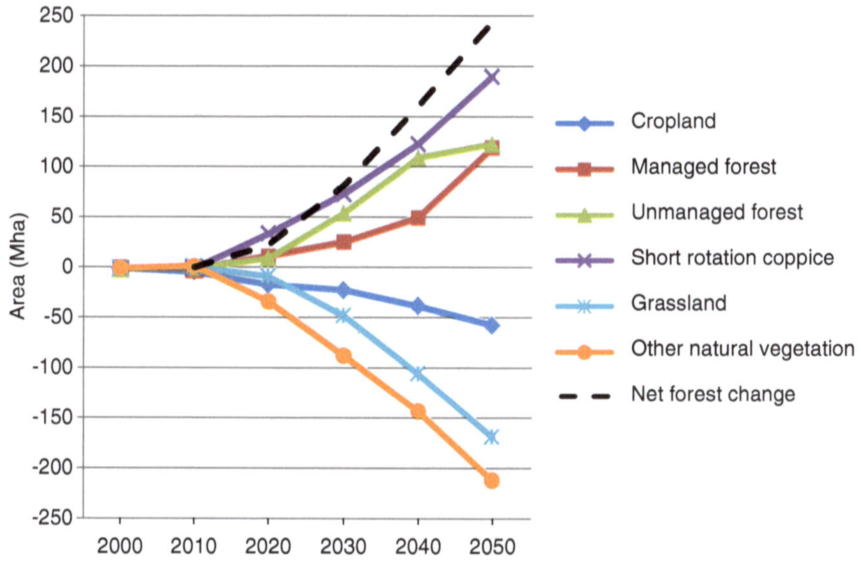

Fig. 5.5 Cumulative land use change and net forest cover change (managed + unmanaged forest area) caused by additional bioenergy production under the *Biodiversity without deforestation* scenario (compared to the 2010 level of bioenergy production) (Kraxner et al. 2013)

5.3.3 Deforestation and Loss of Unmanaged Forest

Deforestation means that unmanaged or managed forest is converted to another land use type, while loss of unmanaged forest implies conversion of unmanaged forest to either managed forest or other land use types. In 2000 the global area of unmanaged forest was 3,146 million ha and the area of managed forest was 719 million ha. In comparison, the area of short rotation plantations, which is not counted as forest land, was 47 million ha.

There is a comparatively small difference in deforestation due to necessary land use change (i.e., change that is not due to land degradation or inefficient land use) between the *Baseline* and the *High bioenergy* scenarios (Fig. 5.6). This can be explained by the fact that by 2050, most options for energy portfolios include huge shares of bioenergy in order to meet the increasing global energy demand (see also comparison of the energy portfolios by Azar et al. [2010]). Even if bioenergy demand is kept at a constant level between 2010 and 2050 (*Low bioenergy* scenario) the accumulated deforested area is as high as 260 million ha in 2050. As there is no restriction on deforestation under the *Low bioenergy* scenario, all bioenergy demand may be sourced from non-protected, pristine forest. The difference in area between *High bioenergy* and the *Low bioenergy* scenarios, 43 million ha, represents the area deforested due to bioenergy production under the *High bioenergy* scenario, which roughly corresponds to the size of Sweden. In contrast, under scenarios with the "no deforestation" target, less than 20 ha of forest will be lost.

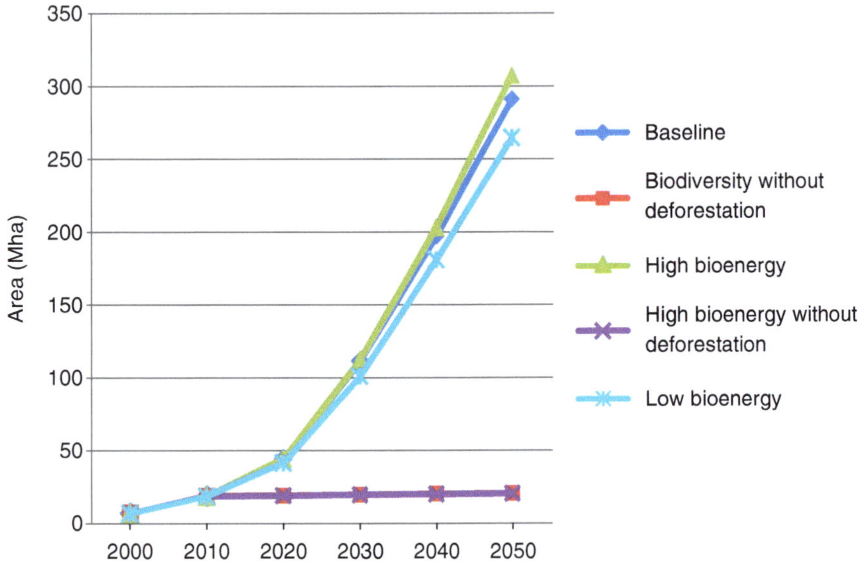

Fig. 5.6 Cumulative deforestation 2000–2050 caused by land use change according to the different scenarios (Kraxner et al. 2013)

Unmanaged forest will be converted to other land use under all scenarios, but under the *High bioenergy without deforestation* and *Biodiversity without deforestation* scenarios the total loss in 2050—336 million ha and 260 million ha, respectively—is only half of the loss under the *Baseline* scenario of 529 million ha. One reason is that intensification in terms of certified sustainable forest management and more intensive agricultural management will help in preserving biodiversity.

Under the *Baseline* scenario, most of the loss of unmanaged forest takes place in the tropical areas of South America, Africa, and Asia. In comparison, under the *High bioenergy without deforestation* scenario, the loss of unmanaged forest is not only considerably smaller but also more evenly distributed from a global perspective. This is also the general pattern under the *Biodiversity without deforestation* scenario. One reason is that a large part of the world's unmanaged forests is located in the tropical areas, and the greatest effects of the "no deforestation" target will thus also be seen in these areas. Nevertheless, after 2040 some conversion of unmanaged forest into managed forest and, to some extent short rotation coppice, takes place under the *High bioenergy without deforestation* scenario in some regions, that is, China, South America, and Pacific Asia. This development is forced by increasing energy demand from a growing world population.

Conversion from unmanaged forest to sustainably managed forest is allowed under all scenarios, and the area of managed forest was also found to increase under all scenarios. The largest increase, of more than 300 million ha, takes place under the *High bioenergy* and *High bioenergy without deforestation* scenarios, primarily in North and South America, China, Pacific Asia and to some extent in Europe. The increase under the *Low bioenergy* scenario is less than half of this; to some extent, the difference can be explained by the difference in bioenergy demand and the fact that under scenarios with the "no deforestation" target, no net deforestation takes place and unmanaged forest is converted into managed forest. Increased future demand for bioenergy will also lead to increase in the area of short rotation plantations, up to a total of between 250 and 300 million ha, in all scenarios but the *Low bioenergy* scenario. The expansion of plantations mainly takes place in the region of former Soviet Union and China, and to a lesser extent in South America and the tropical areas of Africa.

5.3.4 GHG Emissions Due to Total Land Use

The main sources of GHG emissions from land use are deforestation, decay of biomass due to forestry or draining of peatlands, rice cultivation, fertilization, and livestock management, according to the Intergovernmental Panel on Climate Change (Pachauri and Reisinger 2007). In our study, a comparison of the GHG emissions from total land use under the different scenarios shows that bioenergy production indirectly affects GHG emissions considerably through deforestation (Fig. 5.7). Under the *Low bioenergy* scenario, bioenergy use is small compared to the other scenarios, and the GHG emissions are the highest. The GHG emissions are lower under the *Baseline* and *High bioenergy* scenarios, where bioenergy use is more extensive. The GHG emissions are the lowest under the *High bioenergy without*

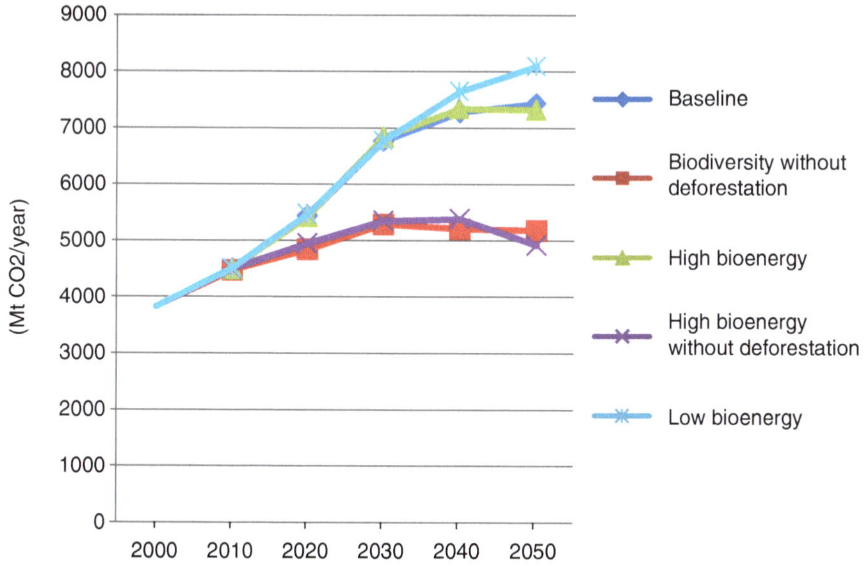

Fig. 5.7 GHG emissions from total land use 2000–2050 under the different scenarios (Kraxner et al. 2013)

deforestation and *Biodiversity without deforestation* scenarios due to the restrictions on deforestation.

GHG emissions from deforestation are highest under the *High bioenergy* scenario, increasing notably between 2020 and 2030. One reason is that unmanaged forest is converted to cropland for bioenergy feedstock production and food production. Another critical factor in terms of GHG emissions is the agricultural sector, that is, soil emissions (N_2O), livestock (CH_4) and rice cultivation (CH_4). However, additional bioenergy production does not change the general pattern of emissions from agriculture compared to the other scenarios. Thus, deforestation can be seen as the largest emission factor, even compared with intensified agricultural practices.

5.4 Implications for Policymaking and Swedish Forestry

According to the scenarios presented here, bioenergy production would not be the major driver of forest loss (see also, Taylor 2011b) but it has significant influence on other potentially negative land use change dynamics. The scenarios indicate that it is possible to avoid large-scale deforestation, even under expanding bioenergy production, by converting unmanaged forests to managed forest. However, note that the scenarios in question assume the forests to be managed according to principles of sustainable forest management. We have to remember that, from a forestry perspective, the 40–50 year outlook provided by the scenarios is a relatively short time span, and unsustainable management with annual harvest levels that are higher than annual growth levels would reduce the biomass potential over time. The projected expansion of managed forests and short rotation plantations is driven primarily by demand for bioenergy. There will be relatively high GHG emissions even under High bioenergy use due to deforestation, agriculture, and natural decomposition processes. However, a deforestation minimization target will help to reduce the GHG emissions substantially.

Overall, the scenarios suggest that with rising populations and projected consumption levels, there will not be enough land to simultaneously conserve all natural areas, stop deforestation, and switch to 100 % renewable energy. Between 2040 and 2050, the food and energy demands of a growing global population will make land competition most acute. The increased pressure on agricultural land implies that management has to become more intense and efficient in terms of irrigation and fertilization, which may in turn lead to high pressure on water resources. This competition for land may lead to rising prices for land as well as for water and food (Schneider et al. 2011; Popp et al. 2012). Impacts on other ecosystems will be greater if forests are more strictly protected. For instance, the loss of other natural vegetation areas is projected to be 263 million ha until 2050 under the *High bioenergy* scenario without deforestation. More than half of that loss is caused by bioenergy production.

In the scenarios presented in this chapter, an increase in the area of natural forest allocated to sustainable management is preferable to outright loss through conversion to energy plantations of forests or other natural ecosystems with a high conservation value. This is based on an assumption that managed forests still will support much of the original biodiversity and ecosystem services (Lattimore et al. 2009). The areas of unmanaged forest with high biodiversity in urgent need of protection are mainly located in the tropics and subtropics. Consequently, extensive areas of boreal and temperate forest will have to be managed to meet a high future demand for bioenergy. Thus, even though pulp and paper industries in South America, China, and southeast Asia are competing with industries in the North, the demand for wood from Swedish forests may be even higher in the future to satisfy demands for sawnwood and bioenergy, especially in a world characterized by concern for environmental issues and climate change (Jonsson 2011). In the light of this, the present debate on the Swedish forestry model is relevant, as it is questioned from both the biodiversity and timber production perspectives. In the Swedish model of forestry, sustainable forest management relies on protection of a smaller proportion of the forest, complemented by general considerations of ecological and social values. General considerations are applied to the whole area of managed forest; for example, trees, groups of trees, buffer zones, and dead wood are retained when harvesting takes place (de Jong et al. 1999). In addition, there are formally and voluntarily protected areas of high natural value which constitute around 3–4 % of the productive forest area (de Jong et al. 1999; Swedish Forest Agency 2007). The concern of the present debate is that general consideration or retention forestry as it is carried out today may not be capable of upholding biodiversity in practice (Gustafsson et al. 2012; Johansson et al. 2009). Furthermore, future demands for wood may further increase the need for more intensive forestry. Zoning of forest land into intensive forestry areas, conventional forestry areas, and protected areas has been proposed as an alternative to the present Swedish model in order to promote increased production and also to protect larger areas of forest (cf. "land sharing" and "land sparing" in Phalan et al. 2011; Ranius and Roberge 2011). The global scenarios presented here underline how important it will be to develop models for forestry that ensure efficient and sustainable forest management both of tropical forests, where the knowledge gaps are very large, and of boreal and temperate forests, in which these models have already been tried and tested in practice.

There will be relatively high GHG emissions even under High bioenergy use due to deforestation, agriculture, and natural decomposition processes. However, a deforestation minimization target will help to reduce GHG emissions substantially. Consequently, well designed mechanisms for reducing emissions from deforestation and forest degradation (REDD) will be crucial for GHG emission reductions and may also contribute to protection of forest biodiversity (Strassburg et al. 2012; Obersteiner et al. 2009). When assessing the relative benefits of bioenergy we also have to consider that sustainable bioenergy is renewable, unlike fossil energy. There is a large potential for carbon sequestration in Swedish forests that may be enhanced with rising temperature from climate change (Böttcher et al. 2011; Briceño-Elizondo et al. 2006). Whereas halting deforestation is the main action needed in the tropics and subtropics, deforestation is not a problem in Sweden. For Sweden (and much of the boreal area) the question is rather how the forests should be managed for optimal climate change mitigation. The optimal strategy depends on system boundaries and what commodities and ecosystem services we wish the forest to provide in addition to carbon sequestration (Backéus 2009). Thus, under a high demand for bioenergy and potentially also a resulting high competition for sawn wood, sustainable management will show a neutral carbon balance if substitution effects are considered (Lippke et al. 2011). Intensive management may increase biomass production and, as a result, carbon sequestration even more (Poudel et al. 2012), but will require strategies for safeguarding other ecosystem services and biodiversity.

The scenarios presented in this chapter are based on global integrated modeling of biomass production and land use. By necessity, certain assumptions and simplifications have to be made as compared to stand-level and landscape-level modeling. Hence, these scenarios should be regarded as projections rather than predictions of the future. It is notoriously difficult to predict what the leading technologies, products, and markets of the future will be. In this study, a large part of our future bioenergy consumption is assumed to be second-generation biofuels, and to ensure that future fuel demands can be satisfied through these, research and development already need to be focusing on technical and practical solutions today, for example, biorefineries (Sims et al. 2010). This calls for innovativeness and flexibility in the forest sector, perhaps especially in countries like Sweden with relatively slow growing forests and high production costs. Moreover, in the future, development of technologies like biomass energy with carbon capture and storage (BECCS) has the potential to make bioenergy carbon-neutral or even carbon-saving (Fuss et al. 2014; Kraxner et al. 2003, 2014; Obersteiner et al. 2001).

We hope that this chapter provides useful examples of the kind of studies needed to guide future policymaking and research. The projections presented in this study regarding trade-offs between deforestation, land use change, and GHG emissions indicate that urgent action with respect to policymaking and good governance is crucial. Renewable energy policies and forest policies should directly address deforestation and forest degradation linked to bioenergy production in order to avoid undesirable effects or even perverse incentives (Searchinger et al. 2010). Various policy areas, inter alia, energy, climate, land use, and rural development, need to be coordinated at all geographic levels and supported by integrated assessments to ensure sustainable use of our common resources.

Acknowledgments The scenarios presented in this chapter are based on scenarios that were first developed jointly with WWF, and they are also presented with additional results and more in-depth analysis in the paper "Global Bioenergy Scenarios—Future Forest Development, Land use Implications and Trade-Offs." We thank the authors of this paper and acknowledge their indirect contribution to this chapter. Further, Eva-Maria Nordström received funding from the Kempe Foundations (Kempestiftelserna) and Future Forests, a multi-disciplinary research program supported by the Foundation for Strategic Environmental Research (MISTRA), the Swedish Forestry Industry, the Swedish University of Agricultural Sciences (SLU), Umeå University, and the Forestry Research Institute of Sweden. We are most grateful for the cooperation with and support from the partner research institutions, mainly the International Institute for Applied Systems Analysis (IIASA) and the World Wide Fund for Nature (WWF). Special gratitude goes to the International Research NGO Global Alpine Synergies (GAS), www.global-syn.org, in Austria for its dedicated support.

References

Azar, C., Lindgren, K., Obersteiner, M., Riahi, K., van Vuuren, D. P., den Elzen, K. M. G. J., Möllersten, K., & Larson, E. D. (2010). The feasibility of low CO_2 concentration targets and the role of bio-energy with carbon capture and storage (BECCS). *Climatic Change, 100*(1), 195–202.

Backéus, S. (2009). *Forest management strategies for CO_2 mitigation.* Doctoral thesis. Acta Univesitatis Agriculturae Sueciae 2009:89. Swedish University of Agricultural Sciences, Umeå.

BirdLife International. (2008). *Endemic bird areas.* Cambridge: BirdLife International.

Böttcher, H., Verkerk, P. J., Gusti, M., Havlik, P., & Grassi, G. (2011). Projection of the future EU forest CO_2 sink as affected by recent bioenergy policies using two advanced forest management models. *GCB Bioenergy, 4*(6), 774–783.

Briceño-Elizondo, E., Garcia-Gonzalo, J., Peltola, H., Matala, J., & Kellomäki, S. (2006). Sensitivity of growth of Scots pine, Norway spruce and silver birch to climate change and forest management in boreal conditions. *Forest Ecology and Management, 232*(1–3), 152–167.

Buongiorno, J., Zhu, S., Zhang, D., Turner, J., & Tomberlin, D. (2003). *The global forest products model.* New York: Elsevier Science.

Capros, P., Mantzos, L., Tasios, N., De Vita, A., & Kouvaritakis, N. (2010). *EU energy trends to 2030 – Update 2009.* Brussels: European Commission Directorate-General for Energy in collaboration with Climate Action DG and Mobility and Transport DG.

Cardellichio, P., Youn, Y., Adams, D., Joo, R., & Chmelik, J. (1989). *A preliminary analysis of timber and timber products production, consumption, trade and prices in the Pacific Rim until 2000* (CINTRAFOR Working Paper 22). Seattle: College of Forest Resources, University of Washington.

de Jong, J., Larsson-Stern, M., & Liedholm, L. (1999). *Grönare skog.* Jönköping: Skogsstyrelsens förlag. [In Swedish]

Duellman, W. E. (1999). *Patterns of distribution of amphibians: A global perspective.* Baltimore: John Hopkins University Press.

EC. (2011). *A roadmap for moving to a competitive low carbon economy in 2050* (Staff Working Document SEC 288). Brussels: European Commission.

FAO. (2006). *World agriculture: Towards 2030/2050. Prospects for food, nutrition, agriculture and major commodity groups.* Rome: Food and Agriculture Organization of the United Nations.

FAO. (2008). *Forests and energy. Key issues* (FAO forestry paper 154). Rome: Food and Agriculture Organization of the United Nations.

Fuss, S., Canadell, J. G., Peters, G. P., Tavoni, M., Andrew, R. M., Ciais, P., Jackson, R. B., Jones, C. D., Kraxner, F., Nakicenovic, N., Le Quéré, C., Raupach, M. R., Sharifi, A., Smith, P., & Yamagata, Y. (2014). Betting on negative emissions. *Nature Climate Change*, 850–853. doi:10.1038/nclimate2392

GEA. (2012). *Global energy assessment – Toward a sustainable future.* Cambridge/New York/Laxenburg: Cambridge University Press/International Institute for Applied Systems Analysis.

Grethe, H., Dembélé, A., & Duman, N. (2011). *How to feed the world's growing billions – Understanding FAO world food projections and their implications.* Berlin: Heinrich Böll Foundation/WWF.

Gustafsson, L., Baker, S. C., Bauhus, J., Beese, W. J., Brodie, A., Kouki, J., Lindenmayer, D. B., Lõhmus, A., Martínez Pastur, G., Messier, C., Neyland, M., Palik, B., Sverdrup-Thygeson, A., Volney, W. J. A., Wayne, A., & Franklin, J. F. (2012). Retention forestry to maintain multifunctional forests: A world perspective. *BioScience, 62*(7), 633–645.

Havlík, P., Schneider, U. A., Schmid, E., Böttcher, H., Fritz, S., Skalský, R., Aoki, K., Cara, S. D., Kindermann, G., Kraxner, F., Leduc, S., McCallum, I., Mosnier, A., Sauer, T., & Obersteiner, M. (2011). Global land-use implications of first and second generation biofuel targets. *Energy Policy, 39*(10), 5690–5702.

Heistermann, M., Müller, C., & Ronneberger, K. (2006). Land in sight? Achievements, deficits and potentials of continental to global scale land-use modeling. *Agriculture, Ecosystems & Environment, 114*(2–4), 141–158.
Hertel, T. W. (1997). *Global trade analysis. Modeling and applications*. Cambridge: Cambridge University Press.
IEA. (2010). *Energy technology perspectives*. Paris: OECD/IEA.
IEA. (2012). *Key world energy statistics*. Paris: OECD/IEA.
Izaurralde, R. C., Williams, J. R., McGill, W. B., Rosenberg, N. J., & Jakas, M. C. Q. (2006). Simulating soil C dynamics with EPIC: Model description and testing against long-term data. *Ecological Modelling, 192*(3–4), 362–384.
Johansson, T., Hjältén, J., de Jong, J., & von Stedingk, H. (2009). *Generell hänsyn och naturvärdesindikatorer – funktionella metoder för att bevara och bedöma biologisk mångfald i skogslandskapet*. Solna: Världsnaturfonden WWF.
Jonsson, R. (2011). Trends and possible future developments in global forest-product markets—Implications for the Swedish forest sector. *Forests, 2*(1), 147–167.
Jull, C., Carmona Redondo, P., Motosi, V., & Vapnek, M. (2007). *Recent trends in the law and policy of bioenergy production, promotion and use* (FAO legislative study 95). Rome: Food and Agriculture Organization of the United Nations.
Kindermann, G. E., McCallum, I., Fritz, S., & Obersteiner, M. (2008). A global forest growing stock, biomass and carbon map based on FAO statistics. *Silva Fennica, 42*(3), 387–396.
Kraxner, F., Nilsson, S., & Obersteiner, M. (2003). Negative emissions from BioEnergy use, carbon capture and sequestration (BECS) – The case of biomass production by sustainable forest management from semi-natural temperate forests. *Biomass and Bioenergy, 24*(4–5), 285–296.
Kraxner, F., Nordström, E.-M., Havlík, P., Gusti, M., Mosnier, A., Frank, S., Valin, H., Fritz, S., McCallum, I., Kindermann, G., See, L., Fuss, S., Khabarov, N., Böttcher, H., Schmid, E., Aoki, K., Máthé, L., & Obersteiner, M. (2013). Global bioenergy scenarios – Future forest development, land-use implications and trade-offs. *Biomass and Bioenergy, 57*, 86–96.
Kraxner, F., Aoki, K., Leduc, S., Kindermann, G., Fuss, S., Yang, J., Yamagata, Y., Il Tak, K., & Obersteiner, M. (2014). BECCS in South Korea – An analysis of negative emissions potential for bioenergy as a mitigation tool. *Renewable Energy, 61*, 102–108.
Lattimore, B., Smith, C. T., Titus, B. D., Stupak, I., & Egnell, G. (2009). Environmental factors in woodfuel production: Opportunities, risks, and criteria and indicators for sustainable practices. *Biomass and Bioenergy, 33*(10), 1321–1342.
Lippke, B., Oneil, E., Harrison, R., Skog, K., Gustavsson, L., & Sathre, R. (2011). Life cycle impacts of forest management and wood utilization on carbon mitigation: Knowns and unknowns. *Carbon Management, 2*(3), 303–333.
Lotze-Campen, H. (2008). The role of modelling tools in Integrated Sustainability Assessment (ISA). *International Journal of Innovation and Sustainable Development, 3*(1–2), 70–92.
Mittermeier, R. A., Gil, P. R., Hoffman, M., Pilgrim, J., & Brooks, T. (2004). *Hotspots revisited: earth's biologically richest and most endangered terrestrial ecoregions*. Mexico City: CEMEX.
Obersteiner, M., Azar, C., Kauppi, P., Möllersten, K., Moreira, J., Nilsson, S., Read, P., Riahi, K., Schlamadinger, B., Yamagata, Y., Yan, J., & Van Ypersele, J. P. (2001). Managing climate risk. *Science, 294*(5543), 786–787.
Obersteiner, M., Huettner, M., Kraxner, F., McCallum, I., Aoki, K., Böttcher, H., Fritz, S., Gusti, M., Havlik, P., Kindermann, G., Rametsteiner, E., & Reyers, B. (2009). On fair, effective and efficient REDD mechanism design. *Carbon Balance and Management, 4*(1), 1–11.
Olson, D. M., & Dinerstein, E. (2002). The global 200: Priority ecoregions for global conservation. *Annals of the Missouri Botanical Garden, 89*(2), 199–224.
Pachauri, R. K., & Reisinger, A. (Eds.). (2007). *Climate change 2007. Synthesis report*. Geneva: IPCC.
Phalan, B., Onial, M., Balmford, A., & Green, R. E. (2011). Reconciling food production and biodiversity conservation: Land sharing and land sparing compared. *Science, 333*(6047), 1289–1291.
Popp, A., Krause, M., Dietrich, J. P., Lotze-Campen, H., Leimbach, M., Beringer, T., & Bauer, N. (2012). Additional CO_2 emissions from land use change—Forest conservation as a precondition

for sustainable production of second generation bioenergy. *Ecological Economics, 74*, 64–70. doi:10.1016/j.ecolecon.2011.11.004.
Poudel, B. C., Sathre, R., Bergh, J., Gustavsson, L., Lundström, A., & Hyvönen, R. (2012). Potential effects of intensive forestry on biomass production and total carbon balance in north-central Sweden. *Environmental Science & Policy, 15*(1), 106–124.
Ranius, T., & Roberge, J.-M. (2011). Effects of intensified forestry on the landscape-scale extinction risk of dead wood dependent species. *Biodiversity and Conservation, 20*(13), 2867–2882.
Ricketts, T. H., Dinerstein, E., Boucher, T., Brooks, T. M., Butchart, S. H. M., Hoffmann, M., Lamoreux, J. F., Morrison, J., Parr, M., Pilgrim, J. D., Rodrigues, A. S. L., Sechrest, W., Wallace, G. E., Berlin, K., Bielby, J., Burgess, N. D., Church, D. R., Cox, N., Knox, D., Loucks, C., Luck, G. W., Master, L. L., Moore, R., Naidoo, R., Ridgely, R., Schatz, G. E., Shire, G., Strand, H., Wettengel, W., & Wikramanayake, E. (2005). Pinpointing and preventing imminent extinctions. *Proceedings of the National Academy of Sciences of the United States of America, 102*(51), 18497–18501.
Russ, P., Wiesenthal, T., van Regemorter, D., & Císcar, J. C. (2007). *Global climate policy scenarios for 2030 and beyond: Analysis of greenhouse gas emission reduction pathway scenarios with the POLES and GEME3 models*. Luxembourg: Joint Research Centre, Institute for Prospective Technological Studies.
Schneider, U. A., Havlík, P., Schmid, E., Valin, H., Mosnier, A., Obersteiner, M., Böttcher, H., Skalský, R., Balkovič, J., Sauer, T., & Fritz, S. (2011). Impacts of population growth, economic development, and technical change on global food production and consumption. *Agricultural Systems, 104*(2), 204–215.
Searchinger, T. D., Hamburg, S. P., Melillo, J., Chameides, W., Havlik, P., Kammen, D. M., Likens, G. E., Obersteiner, M., Oppenheimer, M., Philip Robertson, G., Schlesinger, W. H., Lubowski, R., & David Tilman, G. (2010). Bioenergy: Counting on incentives – Response. *Science, 327*(5970), 1200–1201.
Sims, R. E. H., Mabee, W., Saddler, J. N., & Taylor, M. (2010). An overview of second generation biofuel technologies. *Bioresource Technology, 101*(6), 1570–1580.
Singer, S. (Ed.). (2011). *The energy report: 100 % renewable energy by 2050*. Gland: WWF/Ecofys/OMA.
Strassburg, B. B. N., Rodrigues, A. S. L., Gusti, M., Balmford, A., Fritz, S., Obersteiner, M., Kerry Turner, R., & Brooks, T. M. (2012). Impacts of incentives to reduce emissions from deforestation on global species extinctions. *Nature Climate Change, 2*(5), 350–355.
Stupak, I., Asikainen, A., Jonsell, M., Karltun, E., Lunnan, A., Mizaraite, D., Pasanen, K., Pärn, H., Raulund-Rasmussen, K., Röser, D., Schroeder, M., Varnagiryte, I., Vilkriste, L., Callesen, I., Clarke, N., Gaitnieks, T., Ingerslev, M., Mandre, M., Ozolincius, R., Saarsalmi, A., Armolaitis, K., Helmisaari, H. S., Indriksons, A., Kairiukstis, L., Katzensteiner, K., Kukkola, M., Ots, K., Ravn, H. P., & Tamminen, P. (2007). Sustainable utilisation of forest biomass for energy – Possibilities and problems: Policy, legislation, certification, and recommendations and guidelines in the Nordic, Baltic, and other European countries. *Biomass and Bioenergy, 31*(10), 666–684.
Swedish Forest Agency. (2007). *Fördjupad utvärdering av Levande skogar* (Meddelande 4/2007). Jönköping: Swedish Forest Agency. [In Swedish]
Taylor, R. (Ed.). (2011a). *Chapter 1: Forests for a living planet. WWF living forests report*. Gland: World Wide Fund for Nature.
Taylor, R. (Ed.). (2011b). *Chapter 2: Forests and energy. WWF living forests report*. Gland: World Wide Fund for Nature.
UNEP-WCMC/IUCN WCPA. (2010). *World database on protected areas (WDPA)*. United Nations Environment Programme – World Conservation Monitoring Centre and International Union for Conservation of Nature World Commission on Protected Areas.
Williams, J. R. (1995). The EPIC model. In V. P. Singh (Ed.), *Computer models of watershed hydrology* (pp. 909–1000). Highlands Ranch: Water Resources Publications.
WWF/IUCN. (1994). *Centres of plant diversity: A guide and strategy for their conservation*. Cambridge: World Wide Fund for Nature/International Union for Conservation of Nature.

Chapter 6
Future Forest Governance: Multiple Challenges, Diverging Responses

Katarina Eckerberg

This chapter analyzes the implications of current global trends in forest governance, where increased complexity is leading to the emergence of new conflicts over common-good values between various interests, and to the emergence of new policy instruments and alliances. However, with such multilevel and polycentric policy-making developing in parallel with growing neoliberal economic agendas world-wide, the degree of sustainable management outcomes in forest governance remains highly dependent on strong states and leadership, not least through the European Union to ensure civil society accountability and to counteract asymmetric power relationships.

6.1 Introduction

Forests are key not only to securing the current wood supply and energy needs of the world population, but also to people's aspirations for further economic development. At the same time, there is a need to preserve common-good values such as biological diversity, water resources, and carbon emission sinks. These challenges highlight the importance of examining whether forests are being governed, or should be governed, to meet these multiple goals. This chapter analyzes the implications of current global trends in forest governance, specifically, the interactions between the forest sector and other related policies/sectors and their influence on social and political institutions for forest management in the boreal regions. The main argument is that the institutional governance framework is changing fairly rapidly in the forest sector, leading to the emergence of new conflicts between

K. Eckerberg (✉)
Department of Political Science, Umeå University, Umeå, Sweden
e-mail: katarina.eckerberg@pol.umu.se

E. Westholm et al. (eds.), *The Future Use of Nordic Forests*,
DOI 10.1007/978-3-319-14218-0_6

Photo 6.1 Forests are for future generations

various interests and new alliances. This changing governance framework results from the global trends in demographics, economic growth, land use competition, and climate change policy that has been analyzed in previous chapters. Changes are also related to globalization and affect three main areas, with major implications for forest governance institutions:

(i) The increasing complexity of actors and interests in multilevel and polycentric policymaking processes, including forest industry expansion in terms of territory and sectoral coverage and thus incorporating entire product chains;
(ii) Policy instruments and governance structures that, though moving away from the nation state as the main power center, are still highly dependent on strong states and leadership;
(iii) At the same time as (ii), the growing importance of the European level for forest policy governance.

These three areas of institutional change are explored in this analysis. Governance as a concept has been interpreted in various ways and provides a fluid theoretical basis for empirical research (Rhodes 1996; Pierre 2000; Pierre and Peters 2005). In this chapter, we use "governance" to signify the "erosion of traditional bases of political power," that is, the changing institutional position of the nation state (Pierre 2000, p. 1). This erosion is based on three main simultaneous processes: (i) national governments have, by deregulating financial markets, relinquished control to individual and international actors (e.g., markets and corporations); (ii) non-governmental actors are playing a greater role in policy networks, sometimes regardless of states; and (iii) the positions of local- and regional-level actors are strengthening. When examining the impacts of global forest governance arrangements, we use the framework developed by Bernstein and Cashore (2000), who distinguish "economic globalization" (i.e., the phenomenon of increasing economic integration) from "internationalization" through policymaking. Their analytical framework assumes that domestic policies are shaped by the combined influence of norms and discourses, international rules, markets, and direct access to domestic policy processes. It further highlights the interaction between discursive, economic, and regulatory factors operating internationally and their impacts at the national and local levels. It also stresses the dynamics of regional and local institutional actors functioning in their context-specific environments.

The chapter proceeds as follows. First, I present the expanding landscape of forest policy and show how current global challenges are producing increased institutional complexity. Second, I examine the interaction between polycentric governance levels and the emergence of new policy instruments. Third, I analyze the growing importance of European-level government as it affects the governance structures of the boreal forest region. Finally, I draw conclusions regarding the extent to which, and also how, the institutional framework of forest governance is shifting and the implications of such shifts for fairness and legitimacy, including the various roles that governments can assume in this respect.

6.2 From Dominant Forest Actors to Multiple Actors and Sectors

The diverse and changing demands placed on forests call for appropriate responses from forest policy actors. In addition to the traditional focus on wood, forests have always yielded a broad range of non-timber forest products and services, providing livelihoods for rural communities worldwide. They preserve landscapes and protect soil fertility, and in many areas prevent erosion. They also regulate water flows, reduce flooding, and protect drinking water supplies. More than 20 % of European forests are managed primarily to protect water, soil, and infrastructure, including settlements, roads, railways, pipelines, farmland, and industrial areas (EC 2010). Over time, expectations of forests have increased; forests must not only protect biodiversity and critical habitats but provide major carbon sinks that are crucial in climate change policy. In addition, forests are expected to provide recreational and cultural values for urban and rural dwellers. The mandate for the forestry sector has expanded, which implies the increasing involvement of many other sectors and actors. Over the last 40 years, these shifts have transformed forest policy from a commodity issue into, among other things, a biodiversity, sustainable development, and human rights issue (Arts 2008).

At the same time, competing demands for food and a range of forest products from a growing and generally wealthier world population are putting new pressure on forest resources. In the 1980s papermaking was considered doomed, as new technology was going to replace the need for paper. In reality, paper consumption increased by a factor of 20 in the twentieth century, and more than tripled in the 30 years ending in 1996 (Robins and Roberts 1996, p. 20). In Sweden, for example, paper production increased from 8.4 to 11.4 million tons in the 20 years ending 2010, primarily in the form of exports (Skogsindustrierna 2012). As demand in the developing world continues to grow to meet basic communication and literacy needs, there is as yet little sign of the decoupling of paper consumption from economic growth. China's economic expansion and growth in the wood products industry may signal a different trajectory, should its current wood-saving strategies—in particular, paper recycling—be emulated worldwide (Ajani 2011).

Forests are also universally used as a source of fuel for cooking and heating. Using wood for energy is a common government strategy for curbing global climate change. Bioenergy production is being spurred by European Union policy goals to achieve 20 % renewable energy in the overall energy mix and 10 % in the transport sector by 2020 (COM 2006/848; Directive 2009/28/EC). Moreover, there is continuing pressure to convert forests to food production, which implies that forests are still being cleared for agriculture in many countries. For example, soy plantations in Brazil and palm oil plantations in Malaysia and Indonesia are expanding to meet demand for human consumption and biofuels, with growing Chinese markets acting as an important trigger (Clay 2004).

As a result of diminishing barriers to trade and investment, the Nordic forest-products industry rapidly internationalized in the late twentieth century. Over time, this has also resulted in fewer and larger forest companies. Some of the world's largest and most international forest companies, which have benefited from global economies of scale, though headquartered in the Nordic countries, draw on markets elsewhere. For example, 60 % of the total capacity of the Finnish paper industry is now located outside Finland. Stora Enso, one of the largest private forest companies in the Nordic countries, operates in some 40 countries, and Finnish-based Pöyry has a presence in some 35 countries (Mather 2004). The largest Swedish forest company, SCA, produces in 35 countries and sells in over 100 (SCA 2011). In this sense, the politics of the Nordic countries and global forests are inextricably linked. The links are not limited to the boreal forests, but also connect Nordic companies with developing countries (Lehtinen et al. 2004, p. 256). As suggested by Lehtinen et al. (2004), these Nordic companies could either serve as neocolonialists, disseminating the practices and paradigms of industrial forestry that have become discredited in the North, or they could exert influence to spread more sustainable practices, spurred by nongovernmental organizations (NGOs). Evidence from southeast Asia indicates that pulp and paper manufacturing by large companies in particular has implemented both waste reduction and resource recovery and reuse measures, which are key elements of ecological modernization. However, the third criterion of ecological modernization, namely, dematerialization—or the substitution of high technology for raw material inputs—has not occurred. On the contrary, the expansion of this large-scale industry has resulted in the establishment of pulpwood plantations and the clear-felling of large areas of virgin forests at the expense of rural livelihoods and biodiversity in developing countries. In addition, small- and medium-scale enterprises in these countries, primarily targeting domestic markets, are lagging seriously behind in achieving ecological modernization due to their limited capacity (Sonnenfeld 2000). The lead of Northern forest companies is firmly established, and these companies benefit amply from developing countries' primary forest resources while keeping most of the economic profits of global paper production in the hands of their owners and shareholders. Both Sweden and Finland have enjoyed the benefits of increasing their share of high-value-added forest-based development in the form of printing and writing paper manufacture; at a time of wood supply constraints, they were able to increase their imports of raw material from Russia and the Baltic states as the previously almost closed forest-industrial systems of the former Soviet Union opened up to competition (Lehtinen et al. 2004).

Hence, the nature of the forestry industry and its dependency on national forests has dramatically changed. For example, wood supply in the Nordic countries is increasingly generated from waste paper recycling,[1] which implies that it is becoming more economically viable to locate paper mills near large urban centers rather than near forests and sea transport facilities, as was previously the case. The changing nature of the forest industry, with diversification into higher-technology forest

[1] The recycling rate in Europe reached 64.5 % in 2007, which confirms that the industry is on the path to meeting its voluntary target of 66 % by 2010 (i.e., the ratio of recovered paper utilized for recycling including recovered paper net trade, and paper and board consumption) (European Recovered Paper Council 2007).

products, including a range of packaging, sanitary, and chemical products, and involvement in energy production as well, also suggests that its labor force has become increasingly knowledge-based. Environmental considerations and, to some extent, concern for human rights have become central to the marketing of the Nordic forest industry. To place this shift in context, recall that in the 1980s the forest industry largely considered public concern about the environment a nuisance and disruptive to business as usual (Raitio 2008). Recent developments suggest that the forest industry has been forced to take a broader range of interests into account to avoid being named and shamed for violating environmental and social demands. This does not mean, however, that current forest practices are necessarily more "sustainable," but that more voices are articulating what should be protected and how forests should be managed. Legitimate forest governance hence calls for widening participation and the creation of greater transparency in policymaking processes.

Many of the challenges confronting sustainable forest management worldwide also lie outside the forest sector, namely, in demand for food and agricultural production, energy and biofuels, infrastructure for peri-urban settlements, and measures for climate change mitigation and adaptation. So far, these and many other international and regional forest-related processes have generally failed to generate cross-sectorial communication and collaboration among the many actors interested in these issue areas (McDermott et al. 2010a). The need for a more refined understanding of the dynamics of rapid, interlinked, multiscale social and environmental change has been emphasized, as governance arrangements try to cope with, and adapt to, highly complex and changing environments (Duit et al. 2010). Nevertheless, it has been suggested that new ideas and interpretations concerning sustainability, biodiversity, and governance have now become institutionalized in the field of forest governance processes, generating policy change and innovation (Arts and Buizer 2009). Compared with more classic analyses based on rational choice or purely institutional theory, such discursive–institutional approaches to change foster a more nuanced understanding of global forest policy. This is because changes in discourse are accompanied by coalition (re)formation, changing power relationships between nongovernmental and governmental actors, and new rules of the game over time (Arts and Buizer 2009, p. 341). As will be discussed below, such changes also include the emergence of multilevel governance initiatives in the forest sector.

6.3 Multilevel, Polycentric Policymaking and the Emergence of New Policy Instruments

The above-mentioned emergence of growing and competing demands for food, biofuels, timber, and environmental services severely challenge existing institutions, especially in conjunction with the direct and indirect impacts of climate change. To date, various international processes have developed and complemented each other in attempting to establish key goals and norms for global-scale intergovernmental forest agreements, although few of them are binding (McDermott et al. 2010a). These goals are: (i) to prevent forest loss and promote sustainable forest management through the Forest Principles adopted at the United Nations Conference on Environment and Development (UNCED) in 1992, which also resulted in processes initiated by the United Nations Forum on Forests (UNFF) and the United Nations Framework Convention on Climate Change (UNFCCC); (ii) to combat forest degradation and prepare for climate adaptation through the United Nations Convention to Combat Desertification (UNCCD), the UNFCCC, the Convention on Biological Diversity (CBD), and the UNFF; (iii) to protect biological diversity through the Convention on International Trade in Endangered Species of Wild Flora and Fauna (CITES), the CBD, and the UN Millennium Development Goals (MDGs); (iv) to promote economic development through the General Agreement on Tariffs and Trade (GATT) and the World Trade Organization (WTO), an aspect also mentioned in UNCED, the MDGs, and the UNFF; (v) to produce social welfare and protect human rights through the CBD, the UNFF, the United Nations Declaration on the Rights of Indigenous Peoples (UNDRIP), the International Labour Organization (ILO) Convention No. 169, and the MDGs; and (vi) to improve governance through the 2006 International Tropical Timber Agreement, the CBD, and the Aarhus Convention. However, it should be mentioned that conflicts over how to define a forest, and how to prioritize and trade off social, economic, and environmental values in the quest for sustainable forest management remain to be negotiated and/or resolved in many of these agreements (McDermott et al. 2010a, p. 34). The integration of forests into the international climate regime enables the establishment of a global system of economic incentives tied to emission reductions. For example, the "reducing emissions from deforestation and degradation" (REDD) concept, followed by REDD+ as an offset mechanism within the UNFCCC, provides one of the first sets of rules in international forest governance to have a binding impact on forest practices (Bernstein et al. 2010).

In addition to polycentric policymaking in these parallel and often overlapping international regimes, major trends in contemporary forest governance include the decentralization of forest management, logging concessions in publicly owned commercially valuable forests, and timber certification, primarily in temperate forests (Agrawal et al. 2008). Globally, national governments are by far the largest forest owners, with approximately 86 % of global forests, compared with private ownership of just over 10 % and communal ownership of below 4 % (FAO 2008). However, these figures are misleading with respect to the power relationships in forest governance, as many government-owned forests are managed for multiple purposes by local communities and community-based organizations, while others are managed as private timber concessions by logging companies (Agrawal et al. 2008). An increased number of countries worldwide have devolved certain management decisions in part of their publicly owned forests (notably in protected areas) from central to local government, which means that they currently involve various forms of decentralized resource governance (Andersson and Gibson 2007). Despite the great methodological difficulties in evaluating the impacts of decentralization of this kind on the outcome of forest management regimes, these authors see some evidence that high-performing municipal governance systems in Bolivia have been able to shift activities by local forest users toward land use activities that are less destructive to protected forest areas (Andersson and Gibson 2007, p. 118). In developing countries, decentralization has been promoted by bilateral, multilateral, and private donors and investors who seek improved governance from recipient countries. This has coincided with pressures from local communities and indigenous peoples who want more say in the management of forest resources and to share the benefits accruing from them. In the early 2000s local communities and organizations began governing an additional 200 million hectares relative to the 1980s (Agrawal et al. 2008). More recently, local demands have also been voiced in connection with northern boreal forests, where international networking and the expansion of forest certification standards have been the main drivers of increased concerns for social equity and environmental values (McDermott et al. 2010b), albeit with varying national patterns of participation and impacts at the local level (Keskitalo et al. 2009).

The substantial role of logging companies involved in forest concessions is another factor that diminishes the relative power of national governments in forest management and in bringing market forces more to the fore. Although there are a variety of logging concession arrangements in industrialized countries, where state forest institutions regulate and monitor compliance with forest legislation, private logging companies in the developing world are far less controlled by national governments, leading to an increasing role for commercial forces. The prevalence of illegal logging is an additional threat to sustainable forest practices in those countries which, according to conservative estimates, produce 8–10 % of global wood products (Brack 2003). Recent initiatives to build regional Forest Law Enforcement and Governance (FLEG) agreements constitute an important step in counteracting such problems. They represent a counterforce to the neoliberal economic agendas promoted by many governments, international organizations, and market players, which favor open markets and free trade and lead to massive deforestation (Humphreys 2006). As discussed below, the European Commission is taking a lead role in supporting these regulatory processes.

The debate on certification has emerged as central to the current reconfiguration of the social and environmental credibility of forest-industrial development. Increasing demands for transparency and accountability are now facing forest owners, loggers, and industry in both their forest practices and production processes through various forms of certification and environmental management systems (EMSs). These systems were launched by environmental and social NGOs in collaboration with forest institutions as a way of assuring consumers (largely through retailers) that forest products on the market meet accepted sustainable forest management criteria. To date, however, forest certification has gained ground mainly in temperate forests in the industrialized world, which constitute well over 90 % of the certified forest area worldwide; forests in the developing world are being certified mainly in response to pressure from foreign investors (McDermott et al. 2010b). As

Chan and Pattberg (2008, p. 118) argue, the current geographic patterns of forest certification indicate a bias toward Northern actors that may reinforce and entrench existing asymmetric power relationships over forest resources. Hence, the potential implications of civil society-based accountability as a counterbalance to the increasing accountability gap in global governance may be overstated (Scholte 2004, p. 233). While there are ample examples of environmental NGOs acting as whistle blowers when forest company practices around the world diverge from their stated sustainable forestry policy goals, studies of local communities involved in forest management in Western Europe also suggest a problem with fairness and legitimacy, in that women as well as certain ethnic and user groups are poorly represented (Jeanrenaud 2001).

Not least, fellings of old-growth forests and tropical rain forests are being heavily criticized by environmentalists, scientists, and indigenous peoples, and pressure is mounting on producers and suppliers to safeguard these forests which are valued for their high biodiversity. Assisted by modern communications technology, environmental NGOs increasingly operate across the local, regional, and international scales and share goals. They can exert strong and effective pressure on the operations and policies of individual companies and countries, thereby curtailing the former domination of industry power, particularly in the Nordic setting. However, the scope and impact of NGOs tends to decrease as the distance from Western markets grows (Lehtinen et al. 2004, p. 270).

Global forest governance is also tightly connected with climate politics, as forests play an international role as current and potential carbon sinks. The REDD+ initiative, under the 2009 Copenhagen Accord, provides an international framework for financial support through a number of national partnerships, such as the Forest Investment Program (FIP) of the World Bank and the Global Environment Facility (Kanowski et al. 2011). There are expectations that a focus on REDD+ implementation could also deliver co-benefits, particularly those related to enhancing the capacity and competencies required for institutional and policy reforms in recipient countries, which are essential in addressing "governance gaps" in existing forest policies and in paving the way for potential REDD+ success. Like forest certification, such a development necessitates a "rebalancing of power relationships away from clientelist networks to more pluralistic arrangements involving environmental, community, and indigenous peoples' interests" (Cashore et al. 2006, p. 578). It also assumes that REDD+ arrangements are allowed to emerge from the "bottom," in broad consensus between the government and all relevant stakeholders rather than through an international framework that might impinge upon national sovereignty (Streck 2010). This presupposes the presence of politically legitimate national and subnational settings for forest conservation and management, and that implementation is locally empowering. However, REDD+ processes face the same problems as previous international efforts to protect tropical forests, including weak enforcement, tenure security, and conflict-resolution mechanisms, and have also been criticized for diverting attention away from the protection of biodiversity and neglecting the rights of local communities and indigenous peoples (Kanowski et al. 2011).

6.4 The Increasing Importance of the European Level

Traditionally, the regulatory setting for forestry has varied substantially across national borders, shaped by domestic conditions and the vested interests of governments and the forest industry. Over time, as a result of international cooperation through organizations such as the International Timber and Trade Organization, the European Union, and the North American Free Trade Association (NAFTA), regulation has become more uniform and less country-specific. In Sweden, the somewhat decreased importance of national forest regulations in recent times (with fewer subsidies to forestry and weakened state ownership due to the government's new public management policy) has been compensated for by international re-regulation, including instruments such as environmental management systems (EMSs) and certification systems. In Europe, forest governance is also changing, and one can now speak of the Europeanization of national forest politics, in which national actors in the EU multilevel system of joint decision making affect national actor constellations (Hogl 2000). Despite the absence of a legal foundation for a common forest policy in Europe, there is both functional and cultivated (i.e., political) spill-over in line with integration theory (Haas 1958; Lindberg 1963) through the many actors who are able to affect the content and direction of the European integration process, with those favoring further formal integration having the most influence (Andersson 2007, p. 233). European countries follow a North–South pattern in their attitudes to the Europeanization of forest policy; the southern countries are the most in favor, as they see the protective functions of forests as the prime goal and this is largely the case that has been promoted to date by European agreements (see below). In the forest-dominated countries of northern Europe, where the forest industry has an important economic role, skepticism toward a greater EU role in forest policy has dominated (Andersson 2007, pp. 156–159). However, those patterns are now changing. Swedish economic interest groups, which have an economic and/or industrial interest in forest and forestry, are working as pressure groups to advance integration in close cooperation with EU institutions. Swedish environmental groups are more split, as some fear that forest policy could become even more dominated by economic interests than it is today and that such processes could counteract the influence of emerging global governance instruments to protect environmental interests (Andersson 2007, p. 194).

To date, European regional agreements affecting the forest industry address biodiversity issues through the legally binding Bern Convention to conserve wild fauna and flora and their natural habitats, the EU Natura 2000 network, and the EU Biodiversity Strategy. In addition, forest governance issues are being promoted through the Fourth Ministerial Conference on the Protection of Forests in Europe (now Forest Europe), which commits EU member states to adopting a common approach to national forest program and was supported by an EU Forest Strategy in 1998 and an EU Forest Action Plan in 2006 (Council Resolution 1999/C 56/01; COM 2006/302 final). These initiatives have resulted in a certain streamlining of forest policy across EU member states through both voluntary measures and increased networking among various forest actors (EC 2011). The goal of combating illegal harvesting and illegal timber trade for EU environment and development policies also includes a FLEGT licensing scheme for timber imports from those exporting countries that agree to enter into Voluntary Partnership Agreements with the EU plus an EU regulation prohibiting the sale of illegally harvested timber in the EU. This promotes principles of "good governance," which amount to de facto binding law and also involve civil-society and forest-sector stakeholders in efforts to monitor on-the-ground activities (Bernstein et al. 2010).

It is fair to say, however, that conflicts over European forest policy processes remain. Edwards and Kleinschmit (2012) note three major issues that divide national actors in the European context: (i) subsidiarity and sovereignty versus policy beyond the nation state; (ii) nature conservation versus forest protection; and (iii) UN rules versus independent Forest Europe processes. The Forest Europe process is currently involved in the negotiation of a legally binding agreement for "sustainable forest management" at the pan-European level. However, such initiatives are still contested by those who fear that an agreement might legalize what they consider to be an existing unsustainable solution, given that forest management practices tend to downplay aspects of forestry such as protection of biodiversity and climate adaptation (Dossche and Ozinga 2011). While the conflict between nature conservation and forest production has to do with the concrete issue of how much forest must be protected in order to achieve biodiversity goals, the other two issues concern procedures, that is, the rules of the game. Underpinning subsidiarity and sovereignty interests are issues of national culture and economies, as well as the fact that national actors do not want to give up their control over forests.

6.5 Conclusion: Diverging Forest Governance Pathways

To summarize, the changes taking place globally affect forest governance in the boreal region and the Nordic countries in several ways: (i) through the increasing international role of private forest companies and commercial logging concessions worldwide; (ii) through growing pressure from NGOs, indigenous peoples, and community-based organizations to gain influence over forest management; and (iii) through the rise of international policymaking and new instruments for governing forest resources. Conflicts between forest land uses for producing timber, fuel, food, and a range of other ecosystem services, including those related to climate change, are becoming increasingly apparent as land availability dwindles due to growing populations and consumer demands. In this situation, sustainable forest management is contested ground, as national governments seem to have shrinking powers to protect their forests from commercial exploitation aiming for exclusive profits at the expense of common-good values.

In the Nordic countries, the development of forests and forestry has been relatively sustainable, although this is contested, as the term "sustainability" is loaded with ethical values and subject to ongoing public debate (Beland Lindahl 2008;

Raitio 2008). However, in the near future, environmental values—particularly carbon sequestration—may exceed timber values. This implies that new actors might appear in the policymaking arena, taking what once was unanimity in the forest sector and dividing it into a range of new interest coalitions at multiple levels concerned with, for example, climate change, biodiversity, and food security. Internationally justified demands from indigenous peoples to advance their influence on forest management, on which their traditional culture and livelihood depend, are likely to escalate in coming years, especially in view of the effects of climate change. Climate change impacts are likely to strengthen the above-mentioned governance trends. They will give rise to pressures favoring the greater formalization of policy instruments, as national governments worldwide seek to take advantage of emerging carbon funds at the same time as having to find strategies for dealing with competing demands for food, fuel, and forest products from scarce land resources (Agrawal et al. 2008). This situation will likely intensify in coming decades with the joint effects of climate change, shifting demographic patterns, and generally improved living standards resulting from continued economic growth. In such a situation, national governments could have a shrinking role at a time when the importance of international and local governance is increasing, at least in relative terms.

Nevertheless, an analysis of various pathways of international influence on domestic policies suggests that, while international rules remain weak and nonbinding, parallel influences of norms and discourses, markets, and, in particular, direct access to domestic policy processes are significant in such changes, although the direct causal relationships remain understudied (Bernstein et al. 2010). This influence is exerted through a range of informal policy networks and coalitions, including those of NGOs and educational institutions, which work to reinforce social and environmental values in forest governance. When supplied with sufficient resources and supported by transparency and inclusiveness, national governments still play a crucial role in monitoring and control through national legislative frameworks (Bass and Guéneau 2005). It should be emphasized that the role of national governments in Western countries has been pivotal to the pursuit of sustainable development, as they can intervene in the face of market failure (Baker and Eckerberg 2008). Baker and Eckerberg's (2008) comparative research demonstrates that the state remains a key player in initiating and coordinating sustainable development planning processes: it contributes to capacity building through direct financing, institutional support, and the provision of expertise to subnational authorities; and it initiates and coordinates policy networks and retains considerable power over the nature and functioning of network forms of governance. In forest policy, the state can play an essential role in ensuring that corporations act in the public interest, by harnessing economic power for the benefit of all citizens (Humphreys 2006). Moreover, from a global perspective, sovereign states have maintained their diplomatic role in forging and enforcing international agreements. Governments are also buyers of products and can exert their power through, for example, green procurement for large infrastructure and public housing projects. As governments own the vast majority of forests worldwide, their role in determining the futures of forests should be considerable, as long as they counteract illegal logging and corruption.

The alternative to the neoliberal agenda that currently dominates forest policy processes around the world is the democratic global political will to value all the goods and services that forests provide—public as well as private. As Hoogeveen and Verkooijen (2011) argue, such an alternative does not mean that a new grand instrument needs to be negotiated, but that multiple existing and new initiatives to transform global forest governance must be coordinated. Leadership is essential to such a global endeavor. It is in this spirit that we end this chapter by stating that the future governance of forests must be based on democratic legitimacy in sovereign countries, with transparency and participation by all affected interests in decision-making processes guided by the rule of law. The analysis demonstrates that market-driven governance systems cannot deliver fairness and legitimacy at the global, national, and local levels without strong support from government institutions interacting at multiple levels and with as little Western bias as possible in their exercise of power.

References

Agrawal, A., Chhatre, A., & Hardin, R. (2008). Changing governance of the world's forests. *Science, 320*(5882), 1460–1462.

Ajani, J. (2011). The global wood market, wood resource productivity and price trends: An examination with special attention to China. *Environmental Conservation, 38*(1), 53–63.

Andersson, T. (2007). *En gemensam europeisk skogspolitik? En integrationsteoretisk studie av ett politikområde på tillväxt.* PhD dissertation, Department of Political Science, Umeå University. [in Swedish]

Andersson, K., & Gibson, C. (2007). Decentralized governance and environmental change: How local institutions moderate deforestation in Bolivia. *Journal of Policy Analysis and Management, 26*(1), 99–123.

Arts, B. (2008). *Global governance, NGOs and the politics of scale.* Assen: Van Gorcum.

Arts, B., & Buizer, M. (2009). Forests, discourses, institutions: A discursive–institutional analysis of global forest governance. *Forest Policy and Economics, 11*(5–6), 340–347.

Baker, S., & Eckerberg, K. (2008). *In pursuit of sustainable development: New governance practices at the sub-national level in Europe.* London: Routledge.

Bass, S., & Guéneau, S. (2005). *Global forest governance: Effectiveness, fairness and legitimacy of market-driven approaches* (Idées pour le débat, 13). Paris: Institut du développement durable et des relations internationales (IDDRI).

Beland Lindahl, K. (2008). *Frame analysis, place perceptions and the politics of natural resource management: Exploring a forest controversy in Sweden.* PhD dissertation, Acta Universitatis Agriculturae Sueciae 2008:60 (Swedish University of Agricultural Sciences, Uppsala).

Bernstein, S., & Cashore, B. (2000). Globalization: Four paths of internationalization and domestic policy change: The case of eco-forestry in British Columbia, Canada. *Canadian Journal of Political Science, 33*(1), 67–99.

Bernstein, S., Cashore, B., Eba'a Atyi, R., Maryadi, A., McGinley, K., et al. (2010). Chapter 7: Examination of the influences of global forest governance arrangements at the domestic level. In J. Rayner & B. Katila (Eds.), *Embracing complexity: Meeting global forest governance*

challenges (IUFRO world series), produced by the Expert Panel on the International Forest Regime, the Collaborative Partnership on Forests, Vienna.
Brack, D. (2003). Illegal logging and the illegal trade in forest and timber products. *International Forestry Review, 5*(3), 195–198.
Cashore, B., Gale, F., Meidinger, E., & Newsom, D. (Eds.). (2006). *Confronting sustainability: Forest certification in developing and transition countries*. New Haven: Yale School of Forestry and Environmental Studies.
Chan, S., & Pattberg, P. (2008). Private rule-making and the politics of accountability: Analyzing global forest governance. *Global Environmental Politics, 8*(3), 103–121.
Clay, J. (2004). *World agriculture and the environment*. Washington, DC: Island Press.
COM 2006/302 final: Communication from the Commission to the Council and the European Parliament on an EU Forest Action Plan. Commission of the European Communities, Brussels, Belgium.
COM 2006/848: Renewable Energy Road Map. Commission of the European Communities, Brussels, Belgium.
Council Resolution 1999/C 56/01: Council Resolution of 15 December 1998 on a forestry strategy for the European Union. Official Journal of the European Communities, Brussels, Belgium.
Directive 2009/28/EC "the RES Directive" on the promotion of the use of energy from renewable sources and amending and subsequently repealing Directives 2001/77/EC and 2003/30/EC. Official Journal of the European Union Communities, Brussels, Belgium.
Dossche, V., & Ozinga, S. (2011, June). Legally binding agreement on forests in Europe: "same, but different"? *EU Forest Watch*, Special Issue, FERN, Brussels, Belgium.
Duit, A., Eckerberg, K., Galaz, V., & Ebbesson, J. (2010). Introduction: Governance, complexity, and resilience. *Global Environmental Change, 20*(3), 363–368.
EC. (2010). *Green paper on Forest Protection and Information in the EU*. "Preparing forests for climate change." COM (2010) 66 final, March 2010.
EC. (2011). *Report on the workshop for the review of the EU Forestry Strategy*, 15 April 2011, Brussels. http://ec.europa.eu/agriculture/fore/events/15-04-2011/report_en.pdf. Last accessed 19 Nov 2013.
Edwards, P., & Kleinschmit, D. (2012). Conflicting interests: Discourses and rules in European forest policy, Special Issue on Forest Conflicts. *Forest Policy and Economics*, doi:10.1016/j.forpol.2012.06.002
European Recovered Paper Council. (2007). *European declaration on paper recycling 2006–2010* (Monitoring report 2007). See http://www.intergraf.eu/Content/ContentFolders/PressReleases/2008-09_ERPC_AnnualReport_2007.pdf. Last accessed 19 Nov 2013.
FAO. (2008). *Global forest resource assessment 2005*. Rome: Food and Agriculture Organization of the United Nations.
Haas, E. B. (1958). *The uniting of Europe: Political, social and economic forces 1950–1957*. Stanford: Stanford University Press.
Hoogeveen, H., & Verkooijen, P. (2011). Transforming global forest governance. *Review of Policy Research, 28*(5), 501–508.
Hogl, K. (2000). The Austrian domestic forest policy community in change? Impacts of the globalization and the Europeanisation of forest politics. *Forest Policy and Economics, 1*(1), 3–13.
Humphreys, D. (2006). *Logjam: Deforestation and the crisis of global governance*. London: Earthscan.
Jeanrenaud, S. (2001). *Communities and forest management in Western Europe: A regional profile of the working group on community involvement in forest management*. Gland: Intergovernmental Panel on Forests/IUCN/DFID.
Kanowski, P. J., McDermott, C. L., & Cashore, B. W. (2011). Implementing REDD+: Lessons from analysis of forest governance. *Environmental Science & Policy, 14*, 111–117.
Keskitalo, C., Sandström, C., Tysianouk, M., & Johansson, J. (2009). Local consequences of applying international norms: Differences in the application of forest certification in Northern

Sweden, Northern Finland, and Northwest Russia. *Ecology & Society, 14*(2), 1. [online] http://www.ecologyandsociety.org/vol14/iss2/art1/. Last accessed 19 Nov 2013.
Lehtinen, A. A., Donner-Amnell, J., & Sæther, B. (Eds.). (2004). *Politics of forests: Northern forest-industrial regimes in the age of globalization*. Aldershot: Ashgate.
Lindberg, L. N. (1963). *The political dynamics of European economic integration*. Stanford: Stanford University Press.
Mather, A. (2004). Foreword. In A. Lehtinen et al. (Eds.), *Politics of forests: Northern forest-industrial regimes in the age of globalization* (pp. xi–xiv). Aldershot: Ashgate.
McDermott, C. L., Humphreys, D., Wildburger, C., Wood, P. with regional contributions from Marfo, E., Pacheco, P., & Yasmi, Y. (2010a). Chapter 2: Mapping the core actors and issues defining international governance. In J. Rayner & B. Katila. (Eds.), *Embracing complexity: Meeting global forest governance challenges* (IUFRO world series), produced by the Expert Panel on the International Forest Regime, the Collaborative Partnership on Forests, Vienna.
McDermott, C. L., Cashore, B., & Kanowski, P. (2010b). *Global environmental forest policies: An international comparison*. London: Earthscan.
Pierre, J. (2000). Introduction: Understanding governance. In J. Pierre (Ed.), *Debating governance authority, steering and democracy*. Oxford: Oxford University Press.
Pierre, J., & Peters, G. (2005). *Governing complex societies*. Basingstoke: Palgrave MacMillan.
Raitio, K. (2008) *You can't please everyone: Conflict management practices, frames and institutions in Finnish state forests*. PhD dissertation, Publications in Social Sciences, University of Joensuu, Joensuu.
Rhodes, R. A. W. (1996). The new governance: Governing without government. *Political Studies, XLIV*, 652–667.
Robins, N., & Roberts, S. (1996). *Rethinking paper consumption* (p. 20). London: International Institute for Environment and Development, IIED.
SCA (2011). *SCA webpage*. http://www.sca.com/en/. Last accessed 28 Dec 2011.
Scholte, J. A. (2004). Civil society and democratically accountable global governance. *Governance and Opposition, 39*(2), 211–233.
Skogsindustrierna. (2012). *Webpage* http://www.skogsindustrierna.org/web/Papper_-_produktion_och_leveranser.aspx. Last accessed 14 Jan 2012.
Sonnenfeld, D. (2000). Contradictions of ecological modernization: Pulp and paper manufacturing in South-East Asia. *Environmental Politics, 9*(1), 235–256.
Streck, C. (2010). Reducing emissions for deforestation and forest degradation: National implementation of REDD schemes. *Climatic Change, 100*, 389–394.

Chapter 7
Climate-Related Forest Policies and Trends

Madelene Ostwald

As part of the carbon cycle, forests have a place in climate-related forest policies and trends. By describing forest-related measures driven by international climate negotiations, such as the afforestation and reforestation under the Clean Development Mechanism (CDM) and REDD+ (Reducing Emissions from Deforestation and Forest Degradation), or the voluntary carbon market, this chapter illustrates how carbon has become an important but fuzzy commodity. The demand for carbon-focused measures is also seen in suggested activities in the Swedish context, shown with the Arctic Boreal Climate Development (ABCD) project. It can be said that due perhaps to the complexity involved in quantifying and accounting for carbon, other benefits such as energy substitution or improved hydrology from carbon-improving management strategies are being enhanced in the debate.

7.1 Climate Change and Forestry

There are few areas or sectors that have no connection to ongoing climate change, whether by experiencing its impacts or by contributing to its process. The forest sector is connected to climate change in both respects. The terrestrial system, of which forests are part, is one component of the global carbon cycle that distributes carbon (C) and carbon dioxide (CO_2) among various pools, depending on natural and human influences. Thus, forests became a consideration in the climate change science and debate at an early stage (e.g., IPCC 1990). Forests can be involved in

M. Ostwald (✉)
Centre for Climate Science and Policy Research (CSPR)/Tema V,
Linköping University, Linköping, Sweden

Centre for Environment and Sustainability (GMV),
University of Gothenburg, Göteborg, Sweden
e-mail: madelene.ostwald@liu.se

E. Westholm et al. (eds.), *The Future Use of Nordic Forests*,
DOI 10.1007/978-3-319-14218-0_7

Photo 7.1 Carbon mitigation—a new forest service

mitigation, either as a result of decreased deforestation and deforestation and the corresponding decrease in greenhouse gas (GHG) emissions, or through an increased carbon sink resulting from planting or managing forests to enhance the carbon stock (Sathaye et al. 2001). The world's four billion hectares (ha) of forest (covering 31 % of Earth's total land area) have recently been estimated to contain a carbon stock in above- and belowground biomass of 289 gigatonnes (Gt) of carbon (FAO FRA 2010), corresponding to 45 % of global terrestrial carbon (Malmsheimer et al. 2011). The United Nations Food and Agricultural Organization (FAO) Forest Resource Assessment (FRA) estimated that the amount of carbon stock in forests decreased by 0.5 Gt/year[1] from 2005 to 2010, mainly due to a reduction in the global forest area in Africa, Asia, and South America (Malmsheimer et al. 2011). On the other hand, Pan et al. (2011) used forest inventory data and long-term ecosystem carbon studies to estimate a persistent global carbon sink of 2.4 GtC/year from 1990 to 2007, including a source of emissions from tropical land use changes of 2.9 GtC/year but partially compensated for by tropical regrowth of 1.6 GtC/year. However, estimates of carbon emissions from forests vary depending on the sources and methods used. Van der Werf et al. (2009) highlight this variation, estimating forest GHG emissions at 6 % of total global emissions if uncertainties are included. If uncertainties are disregarded, they suggest 12 %, and if emissions from deforested peatlands (primarily in southeast Asia) are included, the figure is 15 %. The Intergovernmental Panel on Climate Change (IPCC 2007) makes an even higher estimate, stating that forestry GHG emissions contribute 17.4 % of the global total. Despite the great uncertainties in estimating the forest carbon stock and its emissions, forests constitute an important carbon pool and hence an important factor to consider in climate policy processes.

Greenhouse gases and especially carbon have become the dominant units in which all measurements are quantified and counted in international climate politics (Lövbrand and Stripple 2011). This carbon focus can also be observed with regard to forests, where carbon has evolved into an additional "product" alongside well established goods and services such as timber, biodiversity, and ecological services. The increased focus on carbon and its establishment as a major currency can be noted in climate policy-related science assessments such as the year 2000 IPCC Special Report on Land Use, Land Use Change, and Forestry (Watson et al. 2000), which tried to estimate where the carbon is and how it moves. The focus can also be seen in the negotiations within the United Nations Framework Convention on Climate Change (UNFCCC). One outcome of the UNFCCC Conference of the Parties (COP) meeting in December 2011 in Durban on emissions from forests was the decision that the unit to use for quantification was tons of carbon dioxide equivalent per year (tCO_2e; see footnote 1 for a definition) rather than the area of forest in hectares, for example, (UNFCCC 2011).

[1] Total global emissions from all sectors and including all greenhouse gases listed in IPCC's Fourth Assessment Report (2007) are 49 Gt of CO_2 equivalents per year (based on 2004 data); 1 t of C = 3.664 t of CO_2. Due to differences in global warming potential between greenhouse gases, the term CO_2 equivalents (CO_2e) is used.

Although it is tropical forests that have been the focus of international climate policy measures and mechanisms, these international processes also have implications for boreal forests because of their contribution to global GHG emissions through deforestation. This chapter will describe a number of global trends in climate-related forest policies, such as the creation of:

1. The Clean Development Mechanism (CDM);
2. The REDD+ process (i.e., Reducing Emissions from Deforestation and Forest Degradation, plus carbon enhancement, conservation, and sustainable management of forests);
3. The voluntary carbon market; and
4. Climate-related management strategies for multiple benefits.

These processes clearly focus on carbon sequestered or stored in the terrestrial system. Here I will describe how these international policy processes may have influenced actions in the Swedish arena and boreal forest, by citing the example of a fairly new forest management initiative.

7.2 Global Trends in Climate-Related Forest Policy

Despite the considerable uncertainty involved in monitoring, reporting, and verifying (MRV) carbon fluxes and stocks, discussion of the inclusion of terrestrial processes for climate change mitigation has been lively over the last two to three decades (e.g., Alpert et al. 1992; Lasocki 2001). The issue of forest sinks was discussed within the UNFCCC from an early stage, for example, with reference to forest conservation and afforestation measures (UNFCCC 1997). Negotiations have been heated. Consequently, parties to the UNFCCC report their GHG emissions from land use, land-use change, and the forestry sector (LULUCF) to the Climate Change Secretariat as part of their GHG reporting obligations. Annex 1 countries (in essence the developed countries with commitments within the Kyoto Protocol) report annually, while non-Annex I countries (in essence the developing countries) submit their reports less frequently.

The main hurdles to the extensive use of those carbon-sink mechanisms have been associated with methodological issues. These hurdles include baseline development (e.g., what base year or period to use), guaranteeing the additionality of measures taken, guaranteeing long-term carbon storage (permanence), preventing or accounting for leakage, and, not least, funding and costs.

7.3 The Clean Development Mechanism Under the Kyoto Protocol

The idea of conserving forest to reduce GHG emissions and halt climate change was discussed as a component of the Kyoto Protocol but was excluded from it in 2003 (Henders and Ostwald 2012). The forest-related option left in the Kyoto Protocol, an UNFCCC agreement including commitments from 37 industrialized countries that runs from 2008 to 2012, was the afforestation or reforestation (A/R) project type under the Clean Development Mechanism (CDM). CDM projects are to be hosted in developing countries where their GHG emission reductions or savings will generate certified emission reductions (CERs; 1 CER = 1 tCO_2e) saleable on a carbon market. However, due to the high uncertainty, particularly concerning permanence, A/R-produced CERs cannot be traded on the EU Emission Trading System (ETS). Of the approximately 3,900 CDM projects registered in early 2012 covering 15 project types, 36 are A/R projects. Most of the A/R projects have been registered in the last 4 years. Sweden is participating in two of the 36 A/R projects, one in Moldavia and one in Uganda, through the Swedish Energy Agency.

There are only a few A/R CDM projects, generating a small amount of carbon removal in terms of carbon sequestration and they are affecting only a small geographical area. Nevertheless, the process of testing, discussing, and improving the whole portfolio of institutional, methodological, and technical setups for performance-based carbon terrestrial sink policies has had great learning value (see e.g., the IPCC Good Practice Guidance for Land Use, Land-Use Change, and Forestry Penman et al. 2003).

7.4 The Issue of REDD+

Although not yet operational, REDD+, or Reducing Emissions for Deforestation and Forest Degradation[2] in developing countries, has been the central mechanism discussed in climate-related forest policies since the topic was reintroduced in negotiations during the COP 11 climate meeting, held 2005 in Montreal. Two years later at COP 13 in Bali, UNFCCC decided to promote REDD+ through pilot projects and capacity building, which were followed by the frantic acceleration of efforts to understand and reduce the hurdles to such a mechanism. In essence, these are the same methodological hurdles mentioned earlier, namely, baseline setting, additionality, permanence, leakage, and funding. These obstacles are more or less inherent to forest and land-use projects that are required to prove performance.

The basic idea of REDD+ is to create a system that compensates the actors driving deforestation and forest degradation by giving incentives or payments that exceed the gains from alternative land-use practices, thus creating a value for forest conservation. While the REDD+ process has developed since 2007 as a successful component of the climate negotiations, in which it has been perceived as among the most advanced texts supported by a high degree of consensus, it has also been the subject of intense debate and actions by actors outside the Convention. The climate negotiations have largely failed to concretize the future beyond the Kyoto Protocol's first commitment period that ended in 2012, which has also hampered the progress of all smaller components, such as REDD+.

[2] The "+" refers to carbon enhancement, forest conservation, and sustainable management of forests.

Uncertainty over a future climate agreement has to some extent hijacked the slow process of the international negotiations. For that reason, discussions of REDD+ have moved partly outside the UNFCCC policy process, with the support of several actors on the REDD+ pilot scene, such as nations strongly supporting REDD+, for example Norway (Westholm et al. 2011), and programs under the UN and the World Bank. UN-REDD, the Forest Initiative Program (FIP), the Forest Carbon Partnership Facility (FCPF), and the Congo Basin Forest Fund (CBFF) are all examples of REDD+ initiatives at global or regional scales that aim to prepare tropical forest nations for REDD+. Approximately 45 tropical forest nations are in one way or another involved in this early phase of preparation, in which baseline setting, institutional capacity, and tenure issues are key areas of concern (Westholm 2010). The former French President Sarkozy initiated a conference on forests in Paris in March 2010 that summoned high-level representatives from 64 nations focusing on reducing deforestation. The result of the meeting included an additional USD one billion to support REDD+ and the launching of the REDD+ Partnership, which consists of 72 partner countries including both donors and hosts of potential REDD+ projects. Sweden is one of its members and pledged SEK 500 million toward REDD+ in May 2010. Apart from funding already committed to forest-related aid activities, the Swedish Government has also chosen to earmark SEK 100 million for the Global Environmental Facility (GEF). Interest in REDD+ investments peaked in 2009, the same year as the most visited and media-covered COP was held in Copenhagen. According to the self-reporting database of the REDD+ Partnership, the amount of funds pledged by donors exceeded USD 3,000 million in 2009. In 2011, the pledged amount was USD 1,500 million; from a donor's or investor's perspective, storing carbon in standing forest as discussed under UNFCCC auspices might not be a good investment.

7.5 The Voluntary Markets

Due to certain restrictions on the tradability of offsets from land- and forest-based projects, geographical restrictions, and the MRV requirements within the UNFCCC, voluntary markets, which are markets outside the UNFCCC-regulated market, have developed. Here actors can choose to certify their carbon according to standards such as the Verified Carbon Standard (VCS) for trading on markets such as the over-the-counter market and the Chicago Climate Exchange (Forest Trends 2011). Forest-based projects in general have been central to the voluntary markets, accounting for 46 % of market sales in 2010, according to Forest Trends (2011). The largest portion, 29 %, of transactions on the voluntary markets in 2010 occurred in the REDD+/avoided conversion sector (Forest trends 2011) and the first REDD+ credit was issued for a Kenyan carbon project in February 2011. The development of overall market value indicates a sector in great demand: from just USD 8.5 million in 2005, the value of the forest carbon market increased to USD 177.6 million as of 2010 (Diaz et al. 2011). Due to the set-up of the various voluntary market initiatives, the potential to involve the boreal forest is greater in the voluntary markets than in the UNFCCC compliance market.

7.6 Climate-Related Forest Management with Multiple Benefits

As the forest, climate, and carbon debate has proceeded, using REDD+ as the dominant framework over the last half decade, new considerations have emerged (Malmsheimer et al. 2011). The idea of forest management with carbon sequestration as a component has emerged, resulting in ideas for promising win–win situations or even more preferable win–win–win situations. Biological diversity, energy substitution, hazard protection, erosion control, and improved hydrology are goals presented as potential benefits of correct forest management strategy. These goals are often structured by governments into payment for environmental services (PES). Titles such as "Managing forests because carbon matters: Integrating energy, products and land management policy" (Malmsheimer et al. 2011) have appeared in special issues of international scientific journals. Apart from storing carbon or reducing emissions to save the climate, environmental services can be generated and are in demand. These types of environmental services include downstream water management, climate regulation, measures to prevent natural hazards, saved or improved biodiversity, and sustainable livelihoods. With this development, forest policy within climate policy has been a platform or bridge by which new actors and sectors can become involved.

7.7 Carbon for the Sake of the Climate and Boreal Forests

How have all these global carbon stock, emissions, and climate discussions and processes influenced the Swedish forest sector and the boreal forest, which constitutes 18 % of the global forest area? The hottest issue in the UNFCCC climate negotiations, REDD+, excludes boreal forests, while the voluntary market is open to all countries and provides more opportunities for this forest type. As a boreal forest nation, Canada was early in entering the voluntary carbon market with forest-based offsets (Henschel and Gray 2007). As described in the "Transformation of the Canadian forest sector" by Sten Nilsson (Chap. 9), management strategies have undergone great changes in recent decades. This new forest thinking was also accelerated by the timely synchronized discussions of mitigation and adaptation in the light of climate change. In 2008, the Darkwoods Forest Carbon Project was the first Canadian forest carbon project that met the Verified Carbon Standard (VCS) (www.vcsprojectdatabase.org).

In Sweden, one recent sign of the impact of international climate-related forest policy the establishment of the ABCD project in 2009. The ABCD project represents collaboration between the mining company LKAB, one of the largest forest owners in Sweden, Sveaskog, the municipality of Övertorneå, and the county council and administrative board of Norrbotten. The ABCD project aims to reduce the concentration of atmospheric CO_2 and establish a system for trading carbon offsets

(Övertorneå kommun 2011). The two participating companies hope to lead the way to a future offset trading scheme in Sweden and help make Sweden climate neutral by 2025 (Sveaskog 2011). In November 2011, the first payment of SEK 300,000 to Sveaskog was made by LKAB, which estimated that their CO_2 emissions increased by 300,000 tons per year due to increased production (Ny Teknik 2011). The project seeks to improve and increase the forest biomass through better management practices, such as fertilizing the soil with organic carbon and nutrients later in the rotation cycle when the root systems are developed and uptake is improved, to reduce the risk of nutrient leaching. The biomass increase through these carbon sequestering measures will be the additional carbon generated. The earnings will represent the difference between the baseline in a business-as-usual scenario, defined as the minimum requirements stated by the Forestry Act (SVL 2012) for regeneration, management, and harvesting, and the projected baseline under the changed management. The area consists of 50,000 ha, of which Sveaskog owns 40,000 ha and the rest belongs to some 20 private forest owners (Ny Teknik 2011). The idea is to enter or create a carbon market and generate carbon offsets. Apart from increasing the carbon sink in the identified forests, the project also highlights the aims to achieve multiple benefits (Övertorneå kommun 2011). These include:

- Improved forest regeneration
- Increased needle and leaf production over the whole rotation
- Optimal rotation periods
- Gaining experience of creating and handling carbon offsets and the carbon market
- Developing a system for marketing the carbon and offering investments in carbon offsets to private persons and businesses
- Increasing biomass resource for energy production to replace fossil fuels

The project's estimated baseline is that the measures will enhance carbon sequestration by 80,000 tCO_2 per year over and above the business-as-usual scenario. However, as pointed out in the media by representatives of the Swedish Energy Agency, terrestrial sink offsets are ineligible for the regulated EU market (Ny Teknik 2011), so the project owners are strategically directed toward the voluntary markets. The stakeholders' action strategies, which is expanded on in Chap. 8 "Actors' Perceptions and Strategies" by Karin Beland Lindahl is therefore relevant at this stage.

7.8 Concluding Remarks

Outlined broadly, I have tracked the development of carbon as an issue, from being regarded only as part of the global carbon cycle, through a few processes associated with international climate negotiations that are relevant to forestry, and ending in how this development has influenced ideas for boreal forest management. A few conclusions can be drawn that may be relevant to the future:

- Complicated concerns are associated with accounting for forest CO_2 emissions, since they are linked to rewards or penalties. In the cited Swedish case, it will not be possible to trade allowances from the cap-and-trade system under the auspices of UNFCCC and its Kyoto Protocol. This creates the possibility of a voluntary market or an independent offset system. Whatever accounting system is seen in the future of Swedish forest-based carbon-saving initiatives, there is the risk of double accounting, meaning that the same ton of carbon is counted twice, by both the carbon project and Sweden's UNFCCC and Kyoto Protocol GHG inventory. This is a risk, since connections between the different markets, for example, the voluntary market and the Kyoto Protocol market, are often lacking. This has low credibility from a climate efficiency perspective.

- From a biogeochemical perspective, climate change will affect the biomass production of boreal forests, since increased temperatures and growing periods as well as elevated CO_2 concentrations will increase the sink. Estimates indicate increased biomass production in the boreal forest over the next 50–100 years in the order of 12–33 % (Kirilenko and Sedjo 2007; Poudel et al. 2011). This means that the baseline on which a management intervention for carbon offsetting is measured might change in the near future for purely natural reasons and not due to changed practice; hence, the additionality of the intervention can be questioned.
- The form of future international climate commitments, for example whether these will occur under a large-scale umbrella as in the case of the Kyoto Protocol, through more focused and smaller-scale packages (e.g., sinks), or not at all, will influence the forest, climate, and carbon focus of the future.
- Despite the uncertainties concerning international climate commitments under the UN, the voluntary market is flourishing, particularly for forest-based offsets. This indicates that voluntary buyers do have an interest in forest carbon, which means that carbon might continue to be a tradable commodity even without a climate-driven international agreement.
- Forest management strategies including multiple benefits have received increased attention in the scientific literature and in policy processes related to climate initiatives and national strategies. One of these benefits that is often cited is carbon sequestration or storage in the biomass or soil, which is launched as a mitigation measure. The Swedish ABCD project stands as a good example of this multiple strategy, in which carbon, energy substitution, and increased production are seen as products of the same intervention. The idea of generating carbon as a product includes payments based on performance, which in turn require proper MRV and accounting systems to be climate efficient. The outcome of this remains to be followed and evaluated.

Acknowledgments This chapter that was written in early 2013 is based on ideas, discussions, and writings developed jointly with my colleague Lisa Westholm, whom I would like to greatly acknowledge here. I would also like to thank Sabine Henders, Eva Lövbrand, Eskil Mattsson, Reidar Person, and Hans Petersson for their input regarding the text, as well as those involved in the Swedish research network Focali (www.focali.se).

References

Alpert, S. B., Spencer, D. F., & Hidy, G. (1992). Biospheric options for mitigating atmospheric carbon dioxide levels. *Energy Conservation and Management, 33*(5–8), 729–736.
Diaz, D., Hamilton, K., & Johnson, E. (2011). *State of the forest carbon markets 2011*. Washington, DC: Ecosystem Marketplace.
FAO FRA. (2010). *Global forest resource assessment 2010* (FAO forestry paper 163). Rome: Food and Agriculture Organization of the United Nations.
Forest Trends. (2011). *Back to the future state of the voluntary carbon markets 2011*. Washington, DC: Ecosystem Marketplace.
Henders, S., & Ostwald, M. (2012). Forest carbon leakage quantification methods and their suitability for assessing leakage in REDD. *Forests, 3*, 33–58.
Henschel, C., & Gray, T. (2007). *Forest carbon sequestration and avoided emissions*. Kananaskis: Canadian Boreal Initiative/Ivery Foundation.
IPCC. (1990). *Climate change: The IPCC impact assessment 1990*. Canberra: Australian Government Publishing Service.
IPCC. (2007). In Core Writing Team, R. K. Pachauri, & A. Reisinger. (Eds.), *Climate change (2007) synthesis report: Contribution of Working Groups I, II and III to the fourth assessment report of the Intergovernmental Panel on Climate Change*. Geneva: IPCC, 104 pp.
Kirilenko, A., & Sedjo, R. A. (2007). Climate change impacts on forestry. *Proceedings of the National Academy of Science of the United States of America, 104*, 19697–19702.
Lasocki, T. J. (2001). Climate change mitigation through forestry: Theory and practice. *Journal of Sustainable Forestry, 14*, 147–166.
Lövbrand, E., & Stripple, J. (2011). Making climate change governable: Accounting for carbon as sinks credits and personal budgets. *Critical Policy Studies, 5*, 187–200.
Malmsheimer, R. W., Bowyer, J. L., Fried, J. S., Gee, E., Izlar, R. L., Miner, R. A., Munn, I. A., Oneil, E., & Stewart, W. C. (2011). Managing forests because carbon matters: Integrating energy, products and land management policy. *Journal of Forestry, 109*, S5–S50.
Ny Teknik. (2011). *LKABs gröna arbete sågas av EU-regler*. Published 7 December 2011. http://www.nyteknik.se/nyheter/energi_miljo/miljo/article3361669.ece. Last accessed 20 Jan 2012.
Övertorneå kommun. (2011). *ABCD – Arctic Boreal Climate Development*. Retrieved December 02, 2011, from http://www.overtornea.se/naringsliv/Projekt/ABCD---ARCTIC-BOREAL-CLIMATE-DEVELOPMENT/#Delprojekt
Pan, Y., Birdsey, R. A., Fang, J., Houghton, R., Kauppi, P. E., Kurz, W. A., Philips, O. L., Shvidenko, A., Lewis, S. L., Canadell, J. G., Ciais, P., Jackson, R. B., Pacala, S. W., McGuire, A. D., Piao, S., Rautiainen, A., Sitch, S., & Hayes, D. (2011). A large and persistent carbon sink in the world's forests. *Science, 333*, 988–993.
Penman, J., Gytarsky, M., Hiraishi, T., Krug, T., Kruger, D., Pipatti, R., Buendia, L., Miwa, K., Ngara, T., Tanabe, K., & Wagner, F. (2003). *Good practice guidance for land use, land-use change and forestry* (p. 593). Published by the Institute for Global Environmental Strategies (IGES) for the IPCC.
Poudel, B. C., Sathre, R., Gustavsson, L., Bergh, J., Lundström, A., & Hyvönen, R. (2011). Effects of climate change on biomass production and substitution in north-central Sweden. *Biomass and Bioenergy, 35*, 4340–4355.
Sathaye, J. A., Makundi, W. R., Andrasko, K., Boer, R., Ravindranath, N. H., Sudha, P., Rao, S., Lasco, R., Pulhin, F., Masea, O., Ceron, A., Ordones, J., Deying, X., Zhang, X., & Zoumin, S. (2001). Carbon mitigation potential and costs of forestry in Brazil, China, India, Indonesia, Mexico, the Philippines and Tanzania. *Mitigation and Adaptation Strategies for Global Change, 6*, 185–211.
Sveaskog. (2011). LKAB köper koldioxidkrediter av Sveaskog. *Pressmeddelande* www.sveaskog.se/Press-och-nyheter/Pressmeddelanden/2011/LKAB-koper-koldioxidkrediter-av-Sveaskog/. Last accessed 8 Nov 2011.

SVL (2012). Skogsvårdslagstifningen – gällande regler 1 januari 2012. Skogsstyrelsen. ISBN 978-91-88462-92-3.

UNFCCC. (1997). Report on the Subsidiary Body for Scientific and Technological Advice on the work of its fifth session, Bonn, 25–28 February 1997. FCCC/SBSTA/1997/4.

UNFCCC. (2011). Guidance on system for providing information on how safeguards are addressed and respected and modalities relating to forest reference emission levels and forest reference levels as referred to in decisions 1/CP.16. Draft Decision -/CP.17

Van der Werf, G. R., Morton, D. C., DeFries, R. S., Olivier, J. G. J., Kasibhatla, P. S., Jackson, R. B., Collatz, G. J., & Randerson, J. T. (2009). CO2 emissions from forest loss. *Nature Geoscience, 2*, 737–738.

Watson, R., Nobel, I. R., Bolin, B., Ravindranath, N. H., Verardo, D. J., & Dokken, D. J. (2000). *Land use, land-use change and forestry: IPCC special report*. Cambridge: Cambridge University Press.

Westholm, L. (2010). High hopes and fast pace in the REDD+ arena. *Focali Brief* 2010:04, Gothenburg.

Westholm, L., Ostwald, M., Henders, S., & Mattsson, E. (2011). *Learning from Norway – A review of lessons learned for REDD+ Donors* (Focali report no 2011:03). Gothenburg.

Chapter 8
Actors' Perceptions and Strategies: Forests and Pathways to Sustainability

Karin Beland Lindahl

This chapter discusses how the future is handled by actors in the present. It investigates how actors' perceptions of the future—its challenges and its opportunities—influence their strategies and actions. The chapter starts with a frame analysis exploring the visions of a range of actors relevant to Swedish forest sector development. It aims to describe major divisions in the debate on future forest use and on a variety of ways to deal with uncertainty, ambiguity, and ignorance. The analysis relates to international processes important to the Swedish forest sector and feeds into a discussion of competing pathways to "sustainability".

8.1 Politics and Knowledge in the Making

How can we carry out research for a future that is shaped by contingent, contested, and highly politicized governance processes in fields, such as climate change and climate mitigation, in which historical analogies have little direct relevance? One way of tackling this problem is to link history to the future by exploring the drivers of change embedded in the perceptions and strategies of contemporary society. Such an effort is based on the view that the future is socially constructed in the present (see Andersson 2008) and shaped by our visions and statements about it (Sardar 1999).

Frame analysis is typically used to investigate the organization of experience and the action biases to which they give rise. Frames represent people's worlds in ways that call for particular styles of decision making or action (see, e.g., PERRI 6 2005). Divergent frames may give rise to "frame conflicts" and "frame competition"

K. Beland Lindahl (✉)
Unit of Political Science, Luleå University of Technology (LTU), Luleå, Sweden
e-mail: karin.beland.lindahl@ltu.se

E. Westholm et al. (eds.), *The Future Use of Nordic Forests*,
DOI 10.1007/978-3-319-14218-0_8

Photo 8.1 Forest actors hammering out the forest use debate

(Schön and Rein 1994). Indeed, different understandings, or framings, of the same facts, problems, and opportunities are often what political struggles are primarily about (Fischer 2003). In this chapter, frame analysis is used to explore actors' ways of seeing the world, including their perceptions of the challenges and opportunities confronting the forest sector.

Modern society is shaped by new kinds of uncertainties and risks. Beck (2006) argues that it has become a "risk society" in that it is increasingly occupied with debating, preventing, and managing risks that it has itself produced. Risk, according to Beck, is a socially constructed phenomenon. Some people have a greater capacity to define risks than others: "The inequalities of definition enable powerful actors to maximize risks for 'others' and minimize risks for 'themselves'" (p. 333). In the context of global environmental change, incomplete knowledge allows for politicization. Different perceptions of risk, uncertainty, ambivalence, and ignorance become parts of competing visions and strategies. Actors promote frames that support their ideas of "sustainability" and their preferred strategies for achieving it. An effect of this is that the unknown becomes a potentially contested element of actors' competing "pathways to sustainability" (see Beck 2006; Leach et al. 2010). Building on theories of science in society, Jasanoff (2003, p. 227) argues for "technologies of humility" that acknowledge the limits of prediction and control, confront the normative implications of incomplete knowledge, and require forms of citizen participation that go beyond the formal mechanisms. Jasanoff suggests four focal points around which to develop a broader public debate: *framing, vulnerability, distribution,* and *learning*. Together, these provide a useful framework for questions that can be applied to almost every human activity that aims to alter society: "what is the purpose; who will be hurt; who benefits; and how can we know?" (Jasanoff 2003, p. 240).

This book suggests that the global and Swedish forest sectors are in transition. However, the direction of this transition is not very clear. Which pathways will be pursued and which will not is a question of governance and of how emerging pathways interlock with existing institutions and governance processes. The power relations between actors sponsoring conflicting frames and pathways influence which ones are eventually followed—and which will be suppressed.

In the following analysis, the concepts "ecological modernization" and "sustainable development" are used to illuminate the principal differences between actors' response pathways. Some scholars treat ecological modernization as an application of sustainable development (e.g., Dryzek 1997). However, in line with Langhelle (2000), this analysis departs from the view that there are essential differences between the two concepts. In short, ecological modernization is a theory of social change that explores attempts in Northern industrial societies to respond to the negative environmental impacts of modernization (Simonis 1989; Jänicke 1992; Weale 1992; Hajer 1995; 1996; Buttel 2000; Mol and Spaargaren 2000; Spaargaren 2000; Baker 2007). It suggests ways in which environmental problems may be resolved without the course of societal development being completely redirected (Blühdorn 2000; Buttel 2000; Baker 2007), for example, by retaining the notion of rational progress based on economic growth, profit, and consumerism (Blühdorn 2001).

Sustainable development is a disputed concept, with conceptualizations ranging from weak to strong (O'Riordan 1997; Baker 2006a, b). Even so, the original report by the World Commission on Environment and Development (WCED) recognizes that there are ultimate biophysical limits to growth, acknowledges the responsibility of present generations to future generations, challenges the traditional growth paradigm, and addresses questions of distribution, particularly between North and South (WCED 1987; Baker 2006a, 2007). Consequently, a strategy of sustainable development based on the original Brundtland formulation may be seen as a wide-ranging and radical project of social change (Baker 2007).

The frame analysis on which this chapter is based draws primarily on 24 semi-structured interviews, conducted in 2009, with strategically placed actors of relevance to future Swedish forest use. The interviewees include representatives of Nordic forest industry corporations, forest owners in Sweden, a leading international forest consultancy firm, the Swedish Government and state agencies (Ministry of the Environment and the Forest Agency), Swedish and international environmental nongovernmental organizations (E-NGOs), international social movements, the Swedish bioenergy sector, and experts from the Food and Agriculture Organization (FAO), the United Nations Economic Commission for Europe (UNECE), and research organizations such as the International Institute for Applied Systems Analysis (IIASA). The interviews were supplemented with written material such as policy documents, Web sites, and annual reports. Frame analysis was used to explore actors' perceptions and strategies. Three central aspects were focused upon: (1) actors' perceptions of the change facing the forest sector in coming decades; (2) their visions and positions, that is, their perceptions of what is or is not desirable; and (3) their perceptions of and strategies for appropriate action. Frames, including similar perceptions, visions, and strategies, were grouped together, resulting in four groups of internally similar frames (for a more detailed presentation of this procedure, see Beland Lindahl 2008; Beland Lindahl and Westholm 2011).

8.2 A Changing Forest Sector

The forest sector can be defined in broad terms to include "the economic, social and cultural contribution to life and human welfare which is derived from forests and forest-based activities" (Gane 2007, p. 23). It embraces everything to do with forests, including management and governance of the forest resources. Traditionally, the forest sector in most parts of the world has comprised landowners, timber growers, various users of forest land, loggers, transporters, and a range of industries producing pulp, paper, and wood products. The state and its agencies are usually included, as are social and environmental NGOs (see Gane 2007). Following the internationalization of forestry and environmentalism in the 1980s and 1990s, international actors such as the UN and various forest certification organizations have become more prominent. Environmental, and sometimes social, arguments for greater forest conservation have historically made a stand against the forest

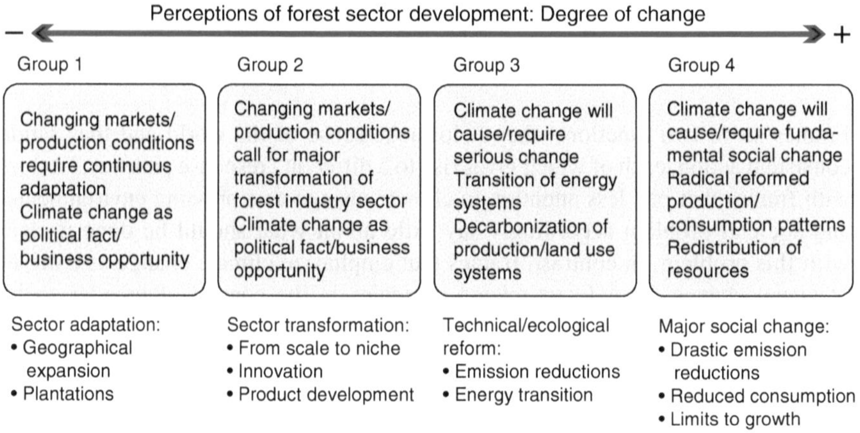

Fig. 8.1 Actors' frames grouped according to their content. Their positions on the continuum indicate diverging perceptions of the scope and degree of change confronting the forest sector

industry's interest of increasing wood production, in Sweden and in many other wood-producing regions.

This situation is now changing. Expected demographic and economic growth, the energy system transition, and the priority given to climate change all affect the structural composition of the forest sector. Bioenergy, carbon sinks, and wood as a substitute for fossil-based materials add to the traditional list of products and services that forests provide. New technologies and practices are therefore being introduced. Changing activities and outputs bring new actors into the sector, while others leave it. The energy industry, together with a range of other climate- and energy-related organizations, are entering the sector with hopes of achieving their aspirations and strategies. The "rules of the game" are changing as forest-related climate and energy policies are introduced. The forest sector is in a state of transition, and its actors are making sense of future challenges and opportunities in various ways. To illustrate this range of understandings and reactions, actors' frames are grouped and organized along a continuum, as shown in Fig. 8.1. The groups should not be seen as fixed and closed; instead, they represent ideal types ranged along a fluid and dynamic continuum.

Two sets of issues reveal more fundamental divides. The first concerns the nature of change, for example, what is *likely* to take place in coming decades. The second emphasizes what is *desirable* and is oriented toward strategies and the action-biasing function of the frame. At the left end of the continuum depicted in Fig. 8.1 are frames in which the scope and transformational strength of future change is relatively limited. These frames are characterized by a relative stability in which the challenges are perceived as manageable. At the right end, in contrast, are frames that envision dramatic changes that will transform both the forest sector and society; these are based more on uncertainty about the nature and scope of interventions needed.

8.3 Actors' Strategies for Coping with Change: Pathways to Sustainability?

Frames serve two functions: they help make sense of the world and they guide consistent action, each of which gives rise to a different "response pathway." Actors with frames that pay less attention to climate change as a pressing environmental and societal problem accordingly say little about what should be done to deal with this problem. In contrast, frames that emphasize climate change as a driver of social change place forest-related strategies in the context of broader social transformation.

Frames in group 1 are found among representatives of Nordic multinational forest industry companies. Group 2 includes frames of other actors from the private Swedish forestry and forest industry sectors, which are not necessarily participating in expanding global markets, and a leading international forestry consultancy firm. In all these frames, climate change-related challenges play a relatively subordinate role. These commercial actors acknowledge their share of emission reductions, but typically explain and justify their commitments and strategies in terms of business opportunities, such as expanding the markets for bioenergy and "renewables" and the prospects of increasing carbon assimilation by boosting forest production. They see climate policies driving a transition to renewable wood-based energy sources as an opportunity. They perceive increased competition from the bioenergy sector and rising energy prices as threats. This ambivalence is reflected in their strategies. At the center of their attention is the forest industry sector and its position in future markets. These actors limit their problem definitions in ways that allow them to deal with problems that are familiar, that is, issues that can be tackled using history, track record, and experience. Accordingly, actors involved in multinational business activities (group 1) stress the continuity and geographical expansion of a relatively traditional forest industry business concept, but now in Latin America and southeast Asia rather than in Europe and North America. Their preferred pathways to sustainability include increased investments in intensive plantation forestry in the South to meet a growing global demand for biomass, contribute to economic development, and assimilate carbon dioxide. Forest industry actors operating in "mature markets" (group 2) focus on what is seen as a necessary restructuring responding to shifts in demand, competition, and access to raw material. Various actors are developing their own strategies for surviving and prospering, but common ideas include moving from "scale" to niche production, intensifying domestic forest production, developing new value chains, products, and services, forming new alliances, and establishing biorefineries. Although reorienting their businesses, actors in groups 1 and 2 do not expand their strategies beyond matters of a relatively familiar character. Therefore, they do not really need to deal with the question of incomplete knowledge and genuine uncertainty or ignorance.

Group 3 includes a range of relatively diverse frames found among actors in various positions in society: UN organizations, such as the Food and Agriculture Organization (FAO) and the United Nations Economic Commission for Europe

(UNECE); the Swedish government, state forestry, and Forest Agency; the bioenergy sector; and more pragmatic E-NGOs, such as the World Wide Fund for Nature (WWF). Group 4 mainly includes the frames of environmental and social NGOs, although components of these frames are shared by some experts affiliated to the FAO and IIASA. Views similar to those of group 4 are also represented by some Southern governments, for example, governments in the West African Economic and Monetary Union (UEMOA). The frames of groups 3 and 4 emphasize climate change as a serious environmental problem that will drive, and entail, significant social change involving the entire forest sector. In contrast to groups 1 and 2, they place problems and strategies in a larger context of climate change that is seen as a major global challenge alongside other environmental problems, such as biodiversity protection. These frames are accordingly biased in favor of strategies and action related to broader ideas of climate mitigation and social change that encompass, and go far beyond, the forest sector. By expanding their problem definitions, these actors open themselves to response pathways that must inevitably deal with greater complexity and incomplete knowledge. However, actors' frames diverge in terms of *how* appropriate pathways ought to be designed.

Group 4, the frames of environmental and social NGOs, is the set of frames that includes the most drastic images of climate- and energy-related social change. Concerns for the environment alongside questions of rural development, resource distribution, and social responsibility are prominent. Logically, social and environmental NGOs upholding these frames advocate actions to drastically reduce emissions, fundamentally change production/consumption systems, reduce consumption levels, and even limit growth in affluent parts of the world. Similar concerns are shared by FAO, UNECE, and UEMOA experts, but the strategies of these organizations focus more on policy and capacity building. The group 4 frames promote sustainability pathways that are open to radical reform of current land use systems, globally and in Sweden. They accept "risk society" in that they base a radical agenda of change on perceptions that threatening future events will occur if action is not taken. They envision a future that is quite different from today's reality, yet without being able to define the exact pathway to get there. However, most of these actors tend to build their strategies and argumentation on components they *know* are crucial, rather than on approaches for dealing with the genuinely unknown.

Most frames in group 3 (e.g., those of the Swedish Government, the Swedish Forest Agency, state forestry, the energy sector, and the WWF) include somewhat less drastic images of future climate/energy-related change. Although organizations such as the WWF argue that we are living beyond our means and call for drastic emission reductions, they focus primarily on positive incentives and "win–win" solutions. Several actors advocate the replacement of fossil-based energy and materials, but their response strategies typically rely on incremental innovation, substitution, increased efficiency, and market-oriented incentives. Compared with group 4, these frames pay less attention to the need to transform production and consumption patterns in more fundamental ways. Rather than facilitating a radical transformation of society, they support various actions to "put things right" and thus restore the idea of continued development along a more or less conventional track.

8.3.1 Strategic Issue 1: The Role of Forest Management

A key question in relation to actors' strategies for coping with climate change is the role of forest management. This is a new issue on the forest political agenda, supported by no historical track record and no consistent science. The state of knowledge is incomplete and various risks are involved, associated with, for example, climate, biodiversity, and livelihoods. What kind of response strategies are actors developing under such conditions?

First, all actors acknowledge that forests have a role to play in climate politics. They all agree that halting deforestation is a priority and that the use of bioenergy, including wood, will increase. However, beyond these commonalities, actors' perceptions diverge widely. Some actors declare that, although forests play a significant role in carbon regulation, strategies to combat climate change should focus on measures outside the forest sector. Consistent with their perceptions of the causes of climate change, the more radical NGOs argue for emission reductions in affluent countries, fear an expansion of industrial plantation forestry in the South, and reject mechanisms to offset forest-related carbon. Experts in the FAO forestry division likewise see measures outside the forest sector as more efficient in terms of climate mitigation and question the wisdom of burning large quantities of wood for energy.

In the frames of other actors, the linkages between climate mitigation, forest management, and wood production are more explicit. However, one important division exists between actors who focus on the carbon *storing* capacity of *standing* forests and those who are preoccupied with the carbon *sequestration* capacity of *growing* forests. Frames of the former type are typically upheld by conservation-oriented NGOs, such as the WWF, and support strategies to make mature forests worth more left standing than felled. They see potential synergies between climate mitigation, forest conservation, and biodiversity protection.

Frames that emphasize the carbon sequestration benefits of younger, growing forests appear in two slightly different versions. One kind, typically upheld by forest owners, industry corporations, and consultants, views managed young forests as carbon sinks and supports actions to intensify forest production, promote plantation forestry, and increase fiber yields. Actors upholding these frames typically perceive the burning of pure wood as waste and argue for further processing before wood and paper products are ultimately burned for energy.

In the other version, climate benefits are primarily associated with the substitution of carbon-neutral wood for fossil-based fuels and materials. Actors upholding these frames typically prioritize the substitution role of wood and come from various camps, for example, the bioenergy sector, the Swedish Forest Agency, and some E-NGOs. The most enthusiastic advocates see forest-based bioenergy as a vital part of climate change mitigation and therefore advocate intensified forest production. Other actors take a more moderate stance. The Swedish Society for Nature Conservation, for example, supports increased use of forest fiber in managing what is perceived as a necessary energy transition, but only to an extent consistent with maintaining biodiversity and recreation values.

With the exception of a few experts, academics, and E-NGO representatives, few actors talk about the difficulties involved in making accurate assessments of forest-related carbon cycling or about incomplete knowledge. As predicted by frame theory, people appear to develop frames and interpret existing knowledge in ways that are consistent with their social commitments and allow them to maintain their ways of life (PERRI 6 2005). Forestry- and forest industry-related actors accordingly emphasize the idea of growing, young forests as *carbon sinks*, as it is biased in favor of action to intensify tree planting and forest production. Actors in the bioenergy sector with a stake in securing their share of the raw material base construct frames in which the climate benefits of substituting *bioenergy for fossil fuels* take precedence. The idea of *standing* forests as *carbon reservoirs* fits well with the traditional conservation strategies of some E-NGOs and is used as an additional argument for *not logging forests*. We can thus see how new issues and trends are conceptualized in ways that are compatible with actors' traditional forest-related frames and strategies of social change. Uncertainty, ignorance, and ambiguity appear to have little room in a highly politicized field in which the actors need solid arguments to maintain their political positions and momentum.

8.3.2 Strategic Issue 2: Biomass Supply and Resource Distribution

Another central issue relates to biomass supply. Is the supply sufficient to meet future needs and how are available resources to be distributed between developing and developed countries, regions, and social groups?

All frames include perceptions of increasing demand for bioenergy and biomass and intensifying competition for raw material. However, actors' frames diverge as to whether it is possible to increase production to meet greater demand or whether shortages are to be expected. A consistent answer to this question is difficult to find in the scientific literature.

Most frames placed in groups 1, 2, and 3 (see Fig. 8.1) lack explicit acknowledgement of constraining biomass shortages of a more permanent nature at the global, or systemic, level. Some actors, such as the WWF, note that our civilization overuses natural resources and expresses concerns about the perceived risk of a demand-driven intensification of land use and forest management. Others, notably the forestry/forest industry and energy sectors, the Swedish Forest Agency, and experts in the FAO forestry division advocate increased forest production and harvesting. Some highlight currently underused resources, others hope for tropical plantation forestry, intensified forest management methods, or the development of new energy technologies. However, few express anxiety at being unable to meet increasing demand. Hence, perceptions of future resource scarcity do not form the primary rationale for strategies and action in most frames in groups 1, 2, and 3. These frames promote response pathways based on ideas of the win–win advantages of institutional and technological changes, emphasizing solutions based on substitution, eco-efficiency, and other measures typical of ecological modernization strategies. Recognizing that they reflect weaker and stronger versions of ecological modernization, these frames all keep relatively quiet on issues of social justice, levels of aggregate resource consumption, ideological conflict, and resource distribution. Implicitly, these frames limit the degree of social and cultural change perceived necessary to address future challenges, especially in high-consumption societies. Strategies of ecological modernization appear to offer sustainability pathways that, though agreeable to actors, are as disparate as forest corporations, ministries, governmental agencies, bioenergy producers, and more pragmatic NGOs.

Frames in group 4, in contrast, include prominent images of resource shortages, increased land use competition, and constraints on increased natural resource exploitation. These frames are found primarily among more radical NGOs, but some experts affiliated with organizations such as the FAO, UNECE, and IIASA foresee similar situations. Perceptions of future competition between North and South, and between food, energy, and forest production, form the basis for their analysis and response strategies. Many of these frames include calls for actions to limit resource exploitation by addressing the demand side, for example, unsustainable consumption patterns in affluent countries. Some recognize limits to growth as an argument for actions to reform the current resource distribution between developed and developing nations. These frames support sustainability pathways requiring more radical social change compatible with the Brundtland version of sustainable development.

Actors who perceive future biomass supply as unlimited, or at least not constraining, see little reason for measures to redistribute available resources. In a situation without tangible shortages, there will be room for everybody and everything. Other actors' calls for fair resource distribution emerge from the perception that there will not be enough for everybody and everything. As Dobson (1998) points out, environmentalism and social justice share a general concern—the existence of scarcity. Environmentalists and advocates of social justice have often chosen to do different things with their conceptions of resource scarcity. However, increasingly shared perceptions of the scarcity of natural resources seem to generate common strategies for reforming both the organization of production and consumption and the distribution of resources and responsibilities. As with the previous issue, actors' traditional forest-related frames and strategies of social change appear important to how they approach new problems about which knowledge is limited and contested.

8.4 Pathways to Sustainability: Tension and Conflict

This analysis suggests that energy, climate, and global land use issues have already entered the forest debate and are likely to become more prominent in the future. As a result, fundamental ideas related to social change, sustainability, and justice are likely to become more salient. A major division exists between frames that promote response pathways based on perceptions of future resource scarcity and those whose strategies are not based on perceived scarcity. The sustainability pathways promoted by these sets of frames may be seen as conflicting rather than complementary, as they are biased in favor of quite different kinds of action. All actors admittedly market their visions and strategies in terms of "sustainable development." Yet supporting pathways based on strategies of ecological modernization envision quite different futures from those promoting a Brundtland version of sustainable development. This difference, or frame conflict, is reflected in actors' forest-related strategies and may fuel future forest debates and conflicts.

As suggested by frame theory, people make sense of new situations in ways that allow them to maintain their social commitments and ways of life (PERRI 6 2005). This analysis suggests that actors' ways of dealing with increasing system complexity and incomplete knowledge follow similar patterns. Some actors, notably the forest industry sector, delimit their problem formulations in ways that enable them to remain on familiar territory where problems are amenable to relatively conventional solutions. A broad range of actors, including the energy sector, Swedish state actors, and pragmatic NGOs such as the WWF, accept an expanded problem formulation, but their response strategies are firmly based on ideas of ecological modernization which assume that the problems can be handled without the existing development trajectory being radically redirected. Drawing on Leach et al. (2010), these strategies may be seen as "equilibrium" responses—"seeking new forms of stable state through a set of interventions, guided by a particular set of knowledge framings, generated by particular practices and institutions" (p. 86). The problem is not that the property of equilibrium, or stability, is necessarily invalid, but that there are powerful pressures that exaggerate its importance and maintain the status quo (see Leach et al. 2010).

Actors with frames promoting more radical social change in line with a Brundtland version of sustainable development, for example, more radical NGOs, tend to be more open to far-reaching transformation that disrupts the existing order of things. Although recognizing that knowledge is limited, most of these actors dwell on issues for which they can construct "fact-based" arguments. In a politicized and polarized debate, little space exists for statements embracing the implications of uncertainty, ambiguity, ignorance, or surprise. Most actors appear to interpret contested or incomplete knowledge selectively, or in ways compatible with their particular vision of social change. This is most evident in their arguments for how various forest management approaches may, or may not, contribute to climate mitigation. In the course of the debate, contested science and partial knowledge are soon converted into "truth." Where knowledge is lacking, the latitude for politicization and struggles over alternative interpretations is the greatest, and where reflection, humility, and precautionary approaches may be needed the most, it appears to be scarce. As Jasanoff (2003) points out, intractable frame conflicts are virtually unavoidable when facts are uncertain.

Which sustainability pathways are—and are not—pursued is largely a question of the power relations between the actors sponsoring the various alternatives. As noted previously, actors tend to construct response pathways that enable them to maintain their social commitments and economic activities. Particular frames give rise to pathways that are promoted by actors with interests in upholding the power of their own institutions and positions to manage the situation. Some pathways, in effect, interlock with powerful parts of the governance system and become self-reinforcing (Leach et al. 2010). Accordingly, pathways promoted by powerful economic interests, state institutions, and influential creators of public opinion are more likely to become dominant than those supported by more marginal interest groups. In this case, powerful commercial and state actors sponsor frames and pathways based on ideas of ecological modernization, that is, frames supporting the status quo or strategies of moderate reform. The actors pushing for more radical social change, headed by the NGOs, generally possess fewer resources and less leverage to exert influence. However, the forest sector is in a stage of transition, and its actors may find reason to reflect and reevaluate their own as well as others' frames. There is also a growing recognition among scientists, governments, and many others of the need to reform the mutual interfaces between policy and science and technology (Jasanoff 2003). Consequently, new alliances, divisions, and power relations may emerge. The rift between actors supporting sustainability pathways promoting the status quo/moderate reform and those promoting radical social change may force the actors to engage in broader debate about the role of forests in future society. Such a development could, for example, exacerbate underlying divisions between mainstream and more radical E-NGOs and encourage progressive parts of forest and bioenergy industries to develop new proactive partnerships addressing the challenges of global change.

As demonstrated here, strong and powerful interests sponsor frames in which problem formulations are delimited in ways that close down the debate and narrow the range of alternative solutions. It is not necessarily in the interests of such actors to expand the problem in such a way that it can encompass broader questions of social change. Indeed, none of the actors in the political arena are tackling the issue of how to deal with ambiguity, complexity, messiness, and questions that have no answers. According to Beck (2006), relationships of definition in risk society are to be conceived as being analogous to Marx's relationships of production. Risk definition is a power game and not all actors benefit from open reflexivity regarding risks and uncertainty. In the present case, uncertainty regarding the optimal use of forests in various pathways to sustainability has been subject to politicization and struggle rather than reflexive deliberation and adjustment. However, it is in the interests of society to avoid lock-in and explore the role of forests in *alternative* pathways to sustainability. The global and Swedish forest sector are in a stage of rapid transition. To the extent that any transition process can be governed or managed, an important aim must be to expand the range of alternatives and to open up the decision space by ensuring that the alternatives now marginalized by dominant actors are heard. Yet, with asymmetric relations between the actors, participation alone is unlikely to promote a meaningful interaction. Posing the questions, "what is the purpose; who will be hurt; who benefits; and how can we know" (Jasanoff 2003, p. 240), could form a starting point for a more reflexive and humble discussion about the roles of forests in alternative pathways to sustainability.

References

Andersson, J. (2008). *The future landscape* (Arbetsrapport 2008:5). Stockholm: Institute for Future Studies.

Baker, S. (2006a). *Sustainable development*. London: Routledge.

Baker, S. (2006b). Environmental values and climate change policy: Contrasting the European Union and the United States. In S. Lucarelli & I. Manners (Eds.), *Values in EU policy* (pp. 77–96). London: Routledge.

Baker, S. (2007). Sustainable development as symbolic commitment: Declaratory politics and the seductive appeal of ecological modernization in the European Union. *Environmental Politics, 16*(2), 297–317.

Beck, U. (2006). Living in the world risk society: A Hobhouse Memorial Public Lecture given on Wednesday 15 February 2006 at the London School of Economics. *Economy and Society, 35*(3), 329–345.

Beland Lindahl, K. (2008). *Frame analysis, place perceptions and the politics of natural Resource Management: Exploring a forest policy controversy in Sweden*. Doctoral dissertation, Swedish University of Agricultural Sciences 2008. Acta Universitatis agriculturae Sueciae 2008:60. Uppsala: Swedish University of Agricultural Sciences.

Beland Lindahl, K., & Westholm, E. (2011). Future forests: Perceptions and strategies of key actors. *Scandinavian Journal of Forest Research, 27*(2), 154–163. doi:10.1080/02827581.2011.635073.

Blühdorn, I. (2000). *Post-ecologist politics: Social theory and the abdication of the ecologist paradigm*. London: Routledge.

Blühdorn, I. (2001). Reflexivity and self-referentiality: On the normative foundations of ecological communication. *Critical Studies, 16*, 181–201.

Buttel, F. H. (2000). Ecological modernisation as a social theory. *Geoforum, 31*, 57–65.

Dobson, A. (1998). *Justice and the environment: Conceptions of environmental sustainability and dimensions of social justice*. Oxford: Oxford University Press.

Dryzek, J. S. (1997). *The politics of the earth: Environmental discourses*. Oxford: Oxford University Press.

Fischer, F. (2003). *Reframing public policy: Discursive politics and deliberative practices*. Oxford: Oxford University Press.

Gane, M. (2007). *Forest strategy: Strategic management and sustainable development for the forest sector*. Dordrecht: Springer.

Hajer, M. A. (1995). *The politics of environmental discourse: Ecological modernisation and the policy process*. Oxford: Oxford University Press.

Hajer, M. A. (1996). Ecological modernisation as cultural politics. In S. Lash, B. Szerszynski, & B. Wynne (Eds.), *Risk, environment and modernity: Towards a new ecology* (pp. 246–268). London: Sage.

Jänicke, M. (1992). Conditions for environmental policy success: An international comparison. *Environmentalist, 12*(1), 47–58.

Jasanoff, S. (2003). Technologies of humility: Citizen participation in governing science. *Minerva, 41*, 223–244.

Langhelle, O. (2000). Why ecological modernization and sustainable development should not be conflated. *Journal of Environmental Policy and Planning, 2*(4), 303–322.

Leach, M., Scoones, I., & Stirling, A. (2010). *Dynamic sustainabilities: Technology, environment, social justice*. London/Washington, DC: Earthscan.

Mol, A., & Spaargaren, G. (2000). Ecological modernisation theory in debate: A review. In A. P. J. Mol & D. A. Sonnenfeld (Eds.), *Ecological modernisation around the world: Perspectives and critical debates* (pp. 17–49). London: Frank Cass.

O'Riordan, T. (1997). Democracy and the sustainability transition. In W. M. Lafferty & J. Meadowcroft (Eds.), *Democracy and the environment: Problems and prospects* (pp. 140–156). Cheltenham: Edward Elgar.

PERRI 6. (2005). What's in a frame? Social organisation, risk perception and the sociology of knowledge. *Journal of Risk Research, 8*(2), 91–118.
Sardar, Z. (1999). The problem of future studies. In Z. Sardar (Ed.), *Rescuing all our futures: The future of futures studies* (pp. 9–19). Twickenham: Adamantine Press.
Schön, D., & Rein, M. (1994). *Frame reflection: Toward the resolution of intractable policy controversies*. New York: Basic Books.
Simonis, U. (1989). Ecological modernisation of industrial society: Three strategic elements. *International Social Science Journal, 121*, 347–361.
Spaargaren, G. (2000). Ecological modernization theory and the changing discourse on environment and modernity. In G. Spaargaren, A. P. J. Mol, & F. H. Buttel (Eds.), *Environment and global modernity* (pp. 41–71). London: Sage.
Weale, A. (1992). *The new politics of pollution*. Manchester: Manchester University Press.
World Commission on Environment and Development (WCED). (1987). *Our common future*. Oxford: Oxford University Press.

Chapter 9
Transition of the Canadian Forest Sector

Sten Nilsson

This chapter examines how Canada is trying to transform the current structure of its forest sector. The Canadian transition is somewhat similar to that of Sweden. Thus, the measures taken by Canada, though differing in many particulars from the Swedish efforts, are also considered as potential "lessons learned" from a Swedish perspective. The forest sectors of the northern hemisphere are undergoing dramatic structural changes as a result of stagnation or decline in the traditional developed economy markets and rapid market growth in the emerging economies. The conditions for the economically sustainable production of industrial forest products are changing rapidly worldwide.

9.1 Introduction

The northern hemisphere forest industry, especially in Canada, is under pressure to transform its practices. At the same time, major forest products such as newsprint (Canada was long the world's leading newsprint producer) and printing- and writing-grade paper are under severe substitution pressure from the increased use of information technologies in the media industry. Concurrently, the forest sectors of the southern hemisphere are improving their competitive position and gaining market share through cheaper production as well as technological change. Today, there are excellent opportunities to produce higher quality paper of most grades using

S. Nilsson (✉)
The Royal Swedish Academy of Agriculture and Forestry,
Stockholm, Sweden

Ecosystems Services and Management Program (ESM),
International Institute for Applied Systems Analysis (IIASA),
Laxenburg, Austria
e-mail: stenbnilsson@gmail.com; nilsson@iiasa.ac.at

E. Westholm et al. (eds.), *The Future Use of Nordic Forests*,
DOI 10.1007/978-3-319-14218-0_9

Photo 9.1 The boreal agreement—going the distance

shorter fibers from fast-growing southern hardwoods instead of longer fibers of northern hemisphere softwoods. The long softwood fibers have long been regarded as a competitive advantage of the northern-based industry; now, however, that advantage is largely gone.

This chapter will examine Canada's efforts to transform its forest sector; possible lessons from the Canadian process for the Swedish transition process will also be discussed.

Canada has approximately 10 % of the world's forests and 30 % of the world's boreal forests. About two-thirds of Canada's estimated 140,000 species of plants, animals, and microorganisms live in its forests. Every year some 1–2 million hectares (ha) of Canada's forests are disturbed by fire and 15–20 million ha are defoliated by insects. The value of the forest-related tourism and recreation industry is assessed to be several billion dollars per year. Canada is also the world's largest forest products exporter, and the yearly average revenue of its forest sector is CAD 50–55 billion.

The need to take Canadian forestry into a new era of global competition and environmental challenges is driven by political and market realities affecting forestry in all boreal regions. Some of the structural developments facing the Canadian forest sector can be summarized as follows.

The Canadian forest sector has been facing a gradually declining wood supply. This decline is caused by the overexploitation of easily accessible forests in many parts of Canada and by tighter environmental restrictions, both of which make wood more expensive at the factory gate. in an industry with a capacity for certain harvest levels, maintaining profitability is difficult once those levels start to decline; forest industry output has also decreased, and certain key products of the Canadian forest industry, such as lumber and newsprint, are facing quickly declining markets. This is due to declining raw material supply and to the changed market conditions and competitiveness position discussed above. Since 1995 capital expenditures in the sector have fallen by some 60 %, meaning that the industry has not maintained its competitive position.

In the 2000s capital costs have substantially exceeded the rate of return on capital employed (ROCE), meaning that the industry has destroyed a lot of capital and now finds it difficult to attract new investment. Declining capital expenditures have resulted in assets with a high technical age; some 60 % of Canada's craft pulp capacity has a technical age of 20 years or more and will probably be too expensive to upgrade. Thus a large portion of the craft pulp capacity will have to be shuttered in coming years. Moreover, mechanical and chemi-thermomechanical pulp mills producing pulp for products such as newsprint and printing- and writing-grade paper are facing declining markets, implying a need for structural change in those subsectors as well.

Most Canadian newsprint production is located in eastern Canada and is high-cost production that needs restructuring. Canada's central/eastern softwood lumber mills are currently among the least cost-effective in the world and also require restructuring. Mehrotra and Kant (2010) published the *Global Competitiveness Index for Forest Product Industries* for six world regions, that is, Canada plus its main competitors. Canada ranked fourth in the softwood lumber sector and sixth in the wood pulp sector, leading the authors to conclude that multi-dimensional interventions are urgently needed in these sectors.

The image of the forest sector in society has become increasingly problematic, and the forest industry and the government have been heavily criticized for shortcomings in forest management and use. There has also been criticism of policies set or not set, how official forest and forest industrial institutions operate, and lack of transparency in decision making.

The above illustrates that the Canadian forest sector is under pressure from changing market conditions rendering the sector less competitive, new southern hemisphere players entering the markets with better pulp and paper products, and technological development favoring the use of fibers from fast-growing hardwood plantations. In all, the economic results of the sector are unsatisfactory and the domestic wood supply is declining.

9.2 Declining Harvests

In Canada, annual allowable cuts (AACs) are set by the provincial governments based on their forestry legislation and taking into account sustainability criteria. The AAC, whose rules and criteria vary from province to province, is not a precise "surgical" instrument, though it does provide an indication of potentially sustainable harvest levels. Over the 1990–2010 period, Canada's total AAC varied between 235 and 250 million m^3/year (National Forestry Database 2011).

Table 9.1 shows Canada's decreasing forest harvests, indicating that the declining activity is not only a result of the 2008–2009 economic downturn but instead a long-term trend. For a few years, the harvests have been substantially below what is regarded as a potentially sustainable level. The actual harvest of 120–200 million m^3/year is substantially lower than the assessed sustainable harvest level of 235–250 million m^3.

Table 9.1 Harvest intensity in the Canadian forest

		m^3 million/year
Allowable annual cut (AAC), 1990–2010 (varying between years)		235–250
Actual harvests	1998	177
	2000	200
	2005	200
	2006	175
	2007	160
	2008	135
	2009	118

Source: Natural Resources Canada (1991–2011)

Several factors have contributed to this declining harvest. In many parts of Canada, areas with good infrastructure have been overharvested, and now companies have to move to more remote areas with less existing infrastructure, resulting in more expensive wood. Greater environmental concern and public pressure demanding sustainable forest management have resulted in political decisions to reduce harvests in several provinces. The changed market conditions for industrial forest products have changed the economic opportunities to use all of the available wood.

The National Forestry Database (2011) presents an outlook for the Canadian forest harvest in 2050, projecting a total decline of some 3 % between 2010 and 2050, though this must be regarded as an underestimate. The forests of British Columbia's interior and, to a lesser extent, Alberta have been attacked by the mountain pine beetle (MPB) in the 2000s. The Government of British Columbia (BC) conducts annual assessments of the forest death caused by the MPB infestation. The latest available assessment is for 2010 (Walton 2011) and encompasses dead forests on Crown Land available for timber production and harvest in BC. The observed killed volume peaked in 2005, since when it has declined gradually. The assessment is that by 2020 the cumulative killed volume will be approximately 850 million m^3, including in private forests (a minor part). The cumulative killed volume in 2010 was assessed at 700 million m^3.

The long-term harvest in BC has averaged approximately 50 million m^3/year, corresponding to 500 million m^3 over a 10-year period. The killed volume over the decade ending 2010 was 700 million m^3 in just one species, the Lodgepole Pine.

There is not enough harvesting and industrial capacity available to process all the killed wood within the short timeframe before it deteriorates, and it is not possible to allocate all the harvest capacity to just the killed wood. From a sustainable forest management perspective, other forests must be harvested as well. The killed wood can be used 2–3 years after the infestation; after that, a strong stain appears in the wood, limiting its applications in industrial processes.

Cullingham et al. (2011) report that the outbreak has affected an accumulated area of 14 million ha in western Canada. The MPB has also moved into Alberta and is attacking Jack Pine, and there is a risk it will continue to move eastwards. This would have strong impacts not only on the AAC of BC but perhaps also on the harvesting potential in Alberta.

So far, there is no consistent picture of the future AAC or harvesting potentials in BC taking the insect outbreak into account. Nilsson (2012) has compiled available information and estimates that the decline of the harvesting potential will start in around 2018 in BC and may reduce harvests by 30–35 % relative to the long-term average harvest of 50 million m^3/year from 1992 to 2010. This depressed harvesting potential will last until at least 2050.

The provinces of Ontario and Quebec have also recently reduced the AAC for political reasons, due to public pressure for more sustainable forest management and management taking greater account of natural values, and because of high logging and transportation costs (in Quebec, the AAC has declined by 25 % in recent years).

A recent National Round Table report (NRT 2011) assesses the impact of climate change on the Canadian forest sector; in total, the harvesting potential may be reduced by 25–35 % in the future. Most of these climate change impacts will take the form of disturbances such as increased wild fires and more serious insect and pest outbreaks.

9.3 Forest Industry in Decline

Canada's industrial forest production is dominated by bulk products, such as lumber, pulp, and newsprint, for export mainly to the United States (USA). Declining harvests are linked to industry problems in a "chicken and egg" way, as decreased production in the industry is in turn reducing the harvest level and harvest declines affect the industry. The export value of forest products more than doubled between 1990 and 2000, followed by a 40 % decline from 2005 to 2010, reflecting severe market problems for several of the bulk products (Natural Resources Canada 1991–2011).

Newsprint production for export declined by 50 % over the last decade, while the export figures fell even more sharply in the face of increased competition from information technology and due to industrial overcapacity and an obsolete newsprint production structure. The drop in newsprint consumption has been most marked in the USA, but has also been substantial in Europe, and conventional wisdom calls for continued decline. Newsprint demand will bottom out one day, but nobody knows at what level or when. Lumber production has also declined, and lumber exports have fallen 50 % despite strongly increased exports to China in 2010. One main reason for the lumber market problems was the collapsing housing market in the USA, resulting in a two-thirds decline in lumber demand between 2006 and 2010. Lumber exports are largely dependent on new housing starts in the USA. There were some 1.6 million housing starts before the economic crisis, and housing starts have remained below 500,000 in recent years. Home construction will pick up again but will not reach the record levels of 1.6 million but perhaps a sustainable rate of one million.

Capital expenditures (i.e., new investments and repair) in the sector, including in the wood, pulp and paper, and logging industries, peaked at CAD 8.7 billion in 1995 and have since dropped to nearly 50 % of that amount. Within 15 years of that peak, capital and repair expenditures have halved and new investments have declined by approximately 60 %. These developments have influenced the profitability and future competitiveness of the Canadian forest industry, as well as its financial performance. In 2000 the industry had an ROCE of nearly 9 % and operating profits of nearly CAD eight billion; over the following decade, however, these figures fell progressively, and financial performance has remained unsatisfactory. In relation to the cost of capital, it is fair to say that the sector in reality destroyed capital during this period: the sector has been unable to attract new investment capital to conduct the required market due diligence or to invest in innovations and new products.

The decline in investment in new facilities and maintenance is attributable to declining or non-existent profitability and to the owners' taking excessive dividends instead of reinvesting. This is a sign that the owners do not really believe in the development opportunities presented by their industry.

The decline of Canada's forest harvest and industry resulted in a reduction of the forest sector's contribution to GDP from 2.8 % in 1995 to 1.9 % by 2010 (Natural Resources Canada 1991–2011) and a loss of altogether 150,000 jobs in the sector over the decade ending 2010. A decline of this magnitude means less interest in the sector from all levels of government.

9.4 Striving for Sustainable Forest Management

Since the 1970s Canadian forestry has been heavily criticized by domestic and international "green" organizations for not carrying out sustainable forest management. Luckert et al. (2011) list some 30 prominent environmental NGOs (E-NGOs) involved in criticizing Canadian forestry in various jurisdictions. The confrontations between E-NGOs, on one hand, and governments and the forest industry, on the other, have periodically been intense. The critics have argued that the industry is mining the Canadian forests by over-harvesting, massive clear-cutting, insufficient reforestation, overuse of environmentally toxic herbicides, destroying wilderness areas, natural forests, and biodiversity, increasing greenhouse gas emissions, damaging and destroying habitats, etc. Critics have also taken aim at institutional and transparency issues with respect to how policies and decisions are made regarding the exploitation and management of Canada's forest resources.[1]

Over the years of criticism, a movement has developed striving for a "new forestry" based on self-sustaining and self-repairing ("resilient") forests, ecology forestry, and sustainable forest management.

Sustainable forest management (SFM) is one approach to sustainable managing forest ecosystems, but one that operates subject only to fairly vague criteria (Duinker et al. 2010). SFM must be monitored in the light of unambiguous criteria that are currently missing in Canada and elsewhere in existing certification systems. Therefore, SFM can only be judged by the progress of the process.

Cashore and McDermott (2004) compared forest policy and regulations across 38 jurisdictions around the world. They state:

> the stringency of Canadian forestry regulations compare with the most stringent of policies elsewhere in the world. In most cases, Canadian policy and regulation is non-discretionary and involves specific requirements. Results of this study indicate that Canada is well positioned to begin a much needed global effort to address important problems concerning the world's forests.

[1] Several leading documents of the criticism of the forest management in Canada are: *Cut & Run: The Assault on Canada's Forests* (Swift 1983), *At the Cutting Edge: The Crisis in Canada's Forests* (May 1998, revised 2005), *Canada's Forests at a Crossroads: An Assessment in the Year 2000* (Global Forest Watch 2000), *The Sacred Balance: Rediscovering our Place in Nature* (Suzuki and McConnell 2002), *The Boreal Forest at Risk: A Progress Report* (Canadian Boreal Initiative 2003), and *The State of Ontario's Forests: A Cause For Concern* (Sierra Club of Canada 2003).

They also reported that Canada has established formal institutional procedures to monitor compliance with sustainable forest management principles. An updated report published in 2010 reached the same positive conclusion (McDermott et al. 2010).

The Library of Parliament (2005) concluded that while there were clear policies and regulations for sustainable forest management established in Canada, implementation was questionable. By 2010 nearly all of the managed forests were certified as sustainably managed (Natural Resources Canada 1991–2011) by the Forest Stewardship Council, Canadian Standards Association, or Sustainable Forestry Initiative.

Despite the progress in increasing the sustainability of forest management, E-NGOs have continued their "Do Not Buy Canadian Wood..." campaigns. Greenpeace convinced some 700 customers of Kimberly-Clark not to buy its products until Kimberly-Clark had agreed to leave non-Forest Stewardship Council-certified boreal forests within 2 years in Canada. In 2009 the agreement was confirmed.

Forest Ethics carried out a campaign against the office supply industry and major customers. Victoria's Secret was one of the companies successfully targeted to stop using paper and paperboard made of pulp originating from the boreal forests in Canada. The organization Canopy focused on the publishing industry and convinced Scholastic Press in 2007 to print *Harry Potter and the Deathly Hallows* only on FSC-certified paper from Canada. This printing was one of the largest single paper orders on record.

Such campaigns have had a great impact on the Canadian forest industry, especially as they target the declining US market for Canadian forest products. Over the first decade of the twenty-first century, the US market fell from receiving 80 down to 62 % of total Canadian forest product exports for the reasons discussed earlier. These campaigns clearly account for some of this shrinkage and have clearly been detrimental to the Canadian forest industry.

9.5 A Needed Transformation

All the conditions described call for the transformation of the Canadian forest sector into something new and different. A few years ago, the Forest Products Association of Canada (FPAC) took on the role of "champion for change." This was possible as the problems were generally recognized but no other strong actors had emerged to drive a transition process. FPAC formulated, with the support of the Canadian Forest Service, what it calls a new strategy addressing four themes:

(i) The competitive position of the Canadian forest sector (assessment of production costs and costs of input factors in various regions of the world),
(ii) New markets in emerging economies,
(iii) Sustainable management of forest resources (addressed by the Canadian Boreal Forest Agreement), and
(iv) New forest products (addressed by the Bio-pathways Project).

The two first themes are fairly traditional and will not be further discussed here; the last two, the Canadian Boreal Forest Agreement and the Bio-pathways Project, are unique and will be examined in greater detail below.

9.5.1 The Canadian Boreal Forest Agreement

The Canadian Boreal Forest Agreement (CBFA; available at www.fpac.ca) was signed in May 2010 by the 21 member companies of FPAC and nine leading environmental organizations in Canada. The agreement, which constitutes a 3-year road-map and mechanisms for extension, applies to 72 million ha of public forests licensed to FPAC members.

The CBFA calls for the suspension of new logging on nearly 29 million ha for the development of conservation plans for endangered boreal (i.e., woodland) caribou, while the fiber supply to the industry is maintained for uninterrupted mill operations. The "Do Not Buy Canadian Wood" campaigns of the environmental organizations have been suspended while the Agreement is being implemented.

The signatories commit themselves to accelerating completion of the protected areas network in the boreal forests, developing and accelerating the implementation of plans to protect species at risk in boreal forests, implementing sustainable forest management based on ecosystem principles, taking action on climate change with respect to forest conservation and product life cycles, taking action to improve the prosperity of the Canadian forest sector, and working to achieve marketplace recognition of environmental performance. The CBFA recognizes that implementation will meet significant challenges. An independent scientific advisory team has been set up to assist signatories in their efforts.

Though the First Nations were not involved in the initial negotiations for the CBFA, the Agreement recognizes that they have constitutionally protected and treaty rights and titles as well as legitimate interests and aspirations. The Agreement states that its signatories will work in accordance with those rights and titles.

The CBFA represents a unique collaboration between industry and environmental organizations working hard to solve common problems through face-to-face exchanges and, if implemented, will conserve significant areas of Canada's boreal forests. Although a collaboration agreement was already reached in 2006 between E-NGOs and the forest industry in the Great Bear Rainforest Agreement applicable to forests on the BC coast, it was on a much smaller geographical scale. International interest in the CBFA has been enormous.

The future of the forests and the environment lie primarily with governments. Both industry and E-NGOs have a responsibility to realize that future, and the CBFA demonstrates the leadership exercised by these parties. The Agreement applies to the 76 million ha under license to FPAC member companies. The agreement—which is not legally binding—covers public forests in BC, Alberta, Manitoba, Saskatchewan, Ontario, and Quebec.

In establishing the CBFA, negotiations were based on "theory U" (Scharmer 2007), which deals with how to solve complex problems when distributed powers are involved. In the end, the theory concerns how to induce interest groups to transcend self-interested positions and interest promotion and to engage in problem solving with perceived opponents.

The basis for the CBFA was described in a speech by FPAC President Lazar as follows:

> The timing for negotiations was right. The NGOs were losing the battle about biodiversity and conservation in spite of successful "Do Not Buy Canadian" campaigns against the forest industry. The industry was losing due to the campaigns, though not that much in sales volumes, but in reputation and damaged brands. Both sides were suffering and wanted to see a solution to the boreal battlefield in Canada. But there were strong psychological constraints for taking the step into negotiations. It meant negotiations with the enemy and there was a strong pressure on persons thinking of going into negotiations not to do so from their own camps. But deep down the parties realized that a status quo situation would not work in the future.... Today, after 3.5 years of negotiations and implementation of the agreement some 70 % of the work within the framework of the agreement is problem solving. A major task is now to sell the results of the problem solving to a broader community outside the partners of the agreement. (http://www.youtube.com/watch?v=43W7MI7UX)

The announcement of the CBFA provoked a mix of passionate reactions in Canada. Some provincial governments, for example, Ontario, expressed strong support for the Agreement. The federal Environment Minster welcomed the Agreement, as did some political party leaders, forest sector unions, and a few First Nations. However, most First Nations were critical because they had not been involved in the process, as were environmental groups excluded from the negotiations (*Dominion* 2010). Somewhat later, the First Nations chiefs demanded that the Agreement be torn up (e.g., *Winnipeg Free Press* 2010; *Working Forest* 2010; *Chronicle Journal* 2011).

The environmental press, including *Nature* (2010), was very positive about the CBFA, describing it as the single largest forest-protection deal ever and as a model of cooperation. A year later, *Nature* (2011) raised concerns as to whether the Agreement would actually succeed: First Nations had been excluded from the deal-making process and were threatening to derail it, industry seemed reluctant to give up logging rights, and environmentalists were demanding greater habitat protection.

In addition to First Nation exclusion from the CBFA process, local communities in the boreal forest were not involved in the process although they were the most affected (*Working Forest* 2010). Some claimed that the 29 million ha of woodland caribou habitat protected in the Agreement would not materialize. Paley (2010) argued that the protected habitat would amount to only a fraction of the claimed 29 million ha, and *Nature* (2011) stated that the area of protected habitat would be under the claimed 29 million ha. The federal government also has some reservations concerning the Agreement because some provinces had not subscribed to it.

To conclude, the CBFA, though initially applauded by many, also has its critics who have noted problems concerning the Agreement's inclusiveness, transparency, accountability, and legal weight.

Photo 9.2 Addressing conflicting interests

The CBFA has also received substantial support from observers following its development over time. The Canadian Institute for Environmental Law and Policy (CIELAP 2011) cited the Agreement as an outstanding example of stakeholder cooperation to resolve conflict. Odendahl (2011), writing in *Listed Magazine*, claimed that the architects of the Agreement probably did the right thing by not trying to please everyone from the outset, concluding that it would have been extremely difficult to get the more than 600 First Nation communities to agree on the framework at the same time. The challenge now facing the Agreement is to reach agreements with the First Nations, provinces, and federal government. As of 2012, this process is under way and memoranda of understanding have been signed between a number of First Nations and CBFA signatories.

Bonoguore (2012) states in a halftime assessment of the 3-year agreement that the "CBFA stands out as a highly innovative and unique approach to a thorny problem." Ben Cashore of the Forest Policy and Governance Program of Yale University, cited in Bonoguore (2012), claims that the CBFA is an excellent example of what can be achieved through a compromise in which E-NGOs must sell wood to uphold conservation values and industry must uphold conservation values to sell wood. However, Cashore would have preferred that all areas covered by the CBFA were FSC certified, which is not the case.

9.5.2 Bio-pathways

The Bio-pathways Project is a major initiative to deal with the poor financial performance of the forest sector, which has been forced to come up with new development paths for economic survival. It was launched in cooperation with FPInnovations (Canada's research organization for forest industrial research and the world's largest such network) and the Ministry of Natural Resources Canada with the objective of investigating the opportunities for industry to produce new products based on wood biomass, products such as biofuels, biochemicals, and biocomposites. Among its specific aims are to help industry identify possible transformational strategies, to create a "sunrise" industry vision, to encourage federal and provincial policymakers to support forest industry transformation, and to provide analytics to investors and partners regarding potential opportunities.

The first phase of the project evaluated established—not laboratory-stage—technologies and real site conditions for production of, for example, bioenergy, biochemicals, and biocomposites from the economic (i.e., ROCE), socioeconomic (i.e., contribution to GDP and employment), and environmental (i.e., GHG footprints) perspectives.

As expected, the economics of the assessed bioproduct technologies varied greatly depending on site conditions, site location, mill configuration, and production scale. Some of the bioproduct technologies displayed much better economic returns than did conventional forest products (e.g., ROCE >20 %). The "best" identified approaches integrate conventional and new technologies, for example, building biorefineries at existing sawmills and pulp mills.

The second phase introduced the market dimension into the analysis and aimed to synthesize information for the bioenergy, biochemical, and biomaterial markets, to develop value chains "from stump to market," to evaluate new business models, and to engage a broad range of stakeholders.

The market potential for new bioproducts was comprehensively analyzed, and a global potential of at least USD 50 billion as of 2015 was identified for green chemicals, wood-fiber composites, and carbon fiber. In addition, strong market opportunities were identified for many other bioproducts. It was concluded that large-volume bioproducts (i.e., commodity types) had the lowest value, while low-volume bioproducts had generally high values per product. This fact alone suggests business models differing greatly from those of the conventional forest industry.

The value chains developed from raw material to market were incremental. For each step in the chains, value-added and ROCE were estimated as were the socioeconomic and environmental impacts. The value chains proved to be extremely important tools for dialog and debate and in engaging stakeholders. Some products displayed excellent potential economic results, which varied between provinces due to varying conditions. Trade-offs between the economic results, socioeconomic impact, and environmental footprint could be identified in several cases. The economically best bioproduct options often had a lower socioeconomic impact and a smaller environmental footprint than did conventional products.

In both Bio-pathways phases, many activities were arranged to involve stakeholders in industrial committees, workshops, government briefings, management briefings in individual forest companies, briefings in the forest industry, and briefings in other sectors, such as the energy, chemical, automotive, finance, and agricultural industries. These activities are still ongoing.

During the Bio-pathways work, it was realized that forming networks with other sectors is crucial. As the assessed forest-based bioproducts will be used in other sectors, the forest industry must understand the functioning of these sectors and their markets. FPAC therefore established the Bio-pathways Partnership Network, a network linking the forest industry with sectors that are potential users of the assessed biomaterials and bioproducts. The purpose of the Network, which now has over 200 members, is to stimulate cooperation between companies from different sectors.

Bio-pathways identified the need to overhaul the overall policy regime in order to avoid policy contradictions and constraints that could hamper the transformation of the forest sector. First, the forest industry, as represented by FPAC, formulated a vision for a transformed forest industry by 2020—"Vision 2020"—which was endorsed in the spring of 2012 (see www.fpac.ca). Second, as the change needed can occur only with the effort of all concerned sectors, FPAC initiated the Bioeconomy Council. This comprises the industrial associations with an interest in bioproducts and the bioeconomy and has the objective of influencing governmental policies affecting the bioeconomy transition. The Bio-pathways Partnership Network and the Bioeconomy Council are two separate bodies with different mandates but similar end goals.

Bio-pathways revealed that, for the transformation of the forest sector, it is not enough simply to develop various new bioproducts. Not all companies are in a position to produce these products, and conventional products will continue to be the backbone of the Canadian forest industry. However, the industry must move toward specialty pulps, nano-cellulose, advanced packaging, high-end hygienic and health products, textiles, and advanced wood products to become economically sustainable. The bioproducts identified in Bio-pathways and value-added conventional products will likely meet and merge; for this reason, FPAC together with the Bio-pathways partners initiated the Value Pathways project in the spring of 2012.

A critical independent review of the Bio-pathways Project in early 2012 demonstrated that the forest industry was still lacking essential components. Willingness to change was in place, but the industry was still defensive regarding the market development of new bioproducts and a market focus was sometimes missing. It is also clear that the forest industry has limited knowledge of the applications and production of consumer products made from new biomaterials. As such knowledge is possessed mainly by sectors outside the forest industry, the forest industry needs frontier partners and alliances to develop products. In addition, the market pull mechanisms are largely missing and must be developed.

The Canadian forest industry has dramatically changed its attitude over the lifespan of the Bio-pathways Project. The industry is now more actively seeking investments in bioproduction, and in fact substantial investments have already been made. The various FPAC initiatives associated with the Bio-pathways Project demonstrate that a sole champion can achieve a lot in a transformation process of the forest sector when the timing is right.

To conclude, the Bio-pathways Project has had major initial impacts, but industry transformation is a long-term process and the full impact of the Project can only be judged further down the road. Bio-pathways offers great opportunities for the Canadian forest industry to include the production of bioproducts. However, that shift will require the joint efforts of the forest industry, government, and industry in other sectors as well as new business models in the forest industry.[2]

9.6 Reflections on Steps Needed for Further Canadian Forest Sector Transition

The following discussion is based on a number of seminars organized by FPAC with critical thinkers from across Canada with respect to forest sector transition.

The CBFA, Bio-pathways, and Value Pathways are important initiatives that are advancing the transition of the Canadian forest sector, but they address only parts of a larger system. To achieve successful transition, the complete system must be dealt

[2] Documentation of all Bio-pathways processes and most of the results of Bio-pathways and Value Pathways can be found at www.fpac.ca.

with. This section will examine some of these other components involved in complete system transformation.

International trends and trajectories, for example, in climate politics, energy markets, and demographics, have a major impact on forest company positioning in Canada. Dialogs with Canadian forest industry actors leave the impression that management in the sector is poorly informed of important global trends and their implications. The main reason for this is the short-term orientation of typical business logics. Governments and public-sector agencies are sometimes more aware of these trends, but take little action since politics also tends to be short-term oriented.

Successful forest sector transformation calls for visions and strategies focused on tomorrow's conditions rather than for reflexive action to protect the existing industry. There is no accepted vision or long-term strategy for the Canadian forest industry or for forest politics. The industry and its owners seem more occupied with acute problems facing individual companies than with general, long-term problems facing the forest industry and sector as a whole. FPAC has, however, formulated its "Vision 2020" applying to the forest industry as comprising its members.

The Canadian government seems little involved in the forest sector. To some extent, the political opposition seems to have identified the problem, stating that the Canadian forest sector is lagging behind those elsewhere in the world due to the lack of a forest sector strategy and of federal government engagement (*Hill Times* 2011).

The program also lacks systemic government policies addressing the forest industry and sector as a whole; instead, existing policymaking has been ad hoc in nature. There is also a need to harmonize policies between provincial and federal governments. Too many policymaking institutions are too narrowly focused to support the transformation of the forest industry and sector. In particular, existing policies prevent a necessary shift from a focus on volume to a focus on value.

Strong leadership is needed in both government and industry to transform the forest sector successfully. Many tough decisions have to be made to facilitate the emergence of a prosperous forest sector. The leadership and champions needed are missing in both government and industry. Strategic thinking lacks diversity and leadership is too risk-averse and too focused on thinking that is short-term and cost-driven, rather than long-term and value-oriented. One explanation for this deficit is the lack of major international forest companies in Canada able to take on leadership roles.

The forest industry has had a difficult time changing its image as a "sunset sector." The industry's bleak financial performance over the last 10 years has made it difficult to generate the substantial investment capital needed for industry renewal. In a situation of limited private capital, governments need to step in and, during the economic crisis, the federal government allocated CAD one billion for the "greening" of the forest industry. In 2010 the federal government introduced the program "Investments in Forest Industry Transformation", which is a 4-year CAD 100 million program. This is far too little money to transform a sector with revenue of some CAD 50 billion/year. This program also lacks a strategic vision and is ad hoc in nature.

R&D and innovations are crucial for the successful transformation of the forest sector. Canada has conducted some high-quality research related to the forest sector, though most of it is defensive in posture. There is a great need for multi-disciplinary, integrated, analytical research teams and for research organizations supporting the development of visions, strategies, and innovations in the forest sector. In the end, forest industry transition is completely dependent on business innovations in the sector. Canada lacks both a national innovation strategy and a focused innovation strategy for the forest sector. An innovation strategy needs to be multi-sector in scope and should provide roadmaps for innovation. Currently, however, the innovation receptor capacity of Canada's forest sector is almost non-existent because poor economic conditions have resulted in very slim organizations and staffing.

Industrial development is dependent on partnerships across industrial sectors. The need for partnerships is rooted in knowledge of new products, markets, distribution networks, business models, leaders, and innovations, so most of the innovations with good industrial potential are being developed outside the forest sector. In addition, partnerships with governments, E-NGOs, and unions need to be developed. The Bio-pathways Partnership Network was established by FPAC in response to this need.

There is an overall "cultural" dimension to industrial renewal in Canada that is difficult to grasp and formulate. Mandel-Campbell (2007) tried to capture this dimension based on interviews with leading Canadians in forest industry, to get to the "soul" of Canadian industry. Canada is one of the richest nations of the world and Canadians are truly privileged. Mandel-Campbell (2007) concludes that Canada has all the prerequisites for becoming a global industrial leader, though she wonders whether Canada wants to become one. She argues that *cultural shifts* are crucial in transforming industry in general, and especially the forest industry and sector, and would turn it in a more prosperous direction. The following quotation captures her view of the culture needed for industrial change in Canada, and in the forest sector:

> Why go through the painful and risky process of building brands, expanding internationally and adding value when there is relatively easy money to be made by carving up home-grown monopolies, cutting down trees and turning out component parts. The Americans phone us and say "we need wood" and we sell it to them and they sell it back to us as cabinets. We are happy to do it because we make money on the wood.

Perhaps the basic question is whether Canada's forest industry and politicians are willing to undertake the necessary transformation of the sector into something more prosperous or whether they will simply stay with an obsolete commodity approach.

Canada is at the frontline of the forest sector transition process by establishing the CBFA, Bio-pathways Project, Value Pathways, Bio-pathways Partnership Network, and Bioeconomy Council. However, dialogs with critical thinkers demonstrate that successful transition calls for more than that. A broad and consistent framework is needed including crucial factors such as leadership, political duration, mentality and attitude, multi-sector thinking, creative innovation environments, and systems thinking. All these are needed to facilitate the concrete implementation of necessary transitional measures.

9.7 The Canadian and Swedish Forest Sectors Compared

Both the Canadian and Swedish forest industries have been characterized by a mature bulk-oriented industrial concept since the early 1980s. A mature industry is reluctant to debate and carry out needed transformations, instead preferring marginal changes to existing business and technological concepts. Ottosson (2011) demonstrates that the most important factors driving the structural change of the Swedish forest industry over the last 20 years have been new energy and climate policies set by government. These are policies that a large part of the Swedish forest industrial collectively opposed for a long time. Canada lacks efficient national policies for energy, climate change, industrial development, and innovation, which hinders the transformation of the forest industry. One example is the establishment of bioenergy production.

In Sweden, over the 2005–2008 period, individual companies became active in driving the transformation of the forest industry with respect to both bioenergy and value-added industrial forest products. This strategy shift occurred because the industry realized the economic potential of bioenergy and the risks of losing this opportunity to the energy industry; in addition, declining markets for traditional forest products (e.g., newsprint and printing- and writing-grade paper) and poor market outlooks for market pulp and traditional lumber made the shift more urgent. Concrete expressions of this strategy shift are Sodra's new bioenergy profile, diversification into higher-value-added industrial products, and reduced dependency on traditional bulk products, the start of SCA's deep transformation from a traditional forest products company into a hygiene products company, Domsjo's expansion into a leading forest biorefinery company, and Billerud-Korsnas' new strategy to become a leading consumer products packaging company. Swedish companies were able to drive this development by themselves because of their solid finances and relatively large international presence in their product segments.

In Canada, in addition to the lack of a conducive policy framework, the forest industry has been financially weak and had too small a presence in international markets—the industry's major focus has always been the US market. Canadian companies lacked the financial strength and probably the managerial "smarts" to undertake the transformation that occurred among individual companies in Sweden. FPAC realized this, and accordingly took the lead in 2009–2010, trying to inspire and motivate the Canadian forest industry to undertake the needed transformational structural changes.

As illustrated earlier, despite some similarities, the Canadian and Swedish forest sectors differ greatly. These differences make it impossible simply to transplant concrete solutions and actions directly from one country to the other with respect to forest sector transition. The important lessons for Sweden from the Canadian transition process in the form of the CBFA and the Bio-pathways Project are the *processes* themselves, which, after some adaptation, can indeed be applied in Sweden.

The CBFA has managed to unite E-NGOs, other interest groups, and industry to work together to solve long-standing problems in the Canadian forest sector.

The agreement has made it possible for actors to move from purely self-interested positions to joint problem-solving. This work was made possible by third-party facilitators throughout the process, which is well organized and follows concrete timetables. The joint problem-solving has led to the abandonment of "we demand" and "we cannot accept" arguments in favor of "the best common solution is." A process of this kind would be interesting to implement in Sweden and would be bound to be a productive problem-solving concept.

Another lesson of the CBFA is that any actor with innovative ideas and truly motivated to find solutions to forest sector problems can do so. The driving force for the agreement was not the government, E-NGOs, or the scientific community but the forest industry lobby organization itself, FPAC. Thanks to innovative thinking and strong management, FPAC managed to drive the individual member companies and sell the concept to E-NGOs and other interest groups. FPAC was an unexpected player driving the process in Canada, and its innovative thinking and strong motivation replaced the slow action of the Canadian government.

A process similar to the Bio-pathways Project does not exist in Sweden. Again, the FPAC served as the driving force for this process too, which underlines the progressive management of the Association. Bio-pathways united the industry within a framework for transition to a new industrial structure by means of new analytical work, backup from the scientific community, support from politicians, government allocation of investment capital for the transition, and the creation of public support for the change. The transition is not only an issue for the conventional forest sector. The new products are linked to and rooted in many other sectors and conventional forest sector players need partners from these sectors if they are to handle the transformation. Bio-pathways organized its Partnership Network that included, as of 2012, some 250 members from various organizations, with regular meetings organized by the project. The Network's partners have shared interests in new products and are starting to work together.

In addition, many policy bottlenecks impede the transition to more sustainable, "greener" industrial production. These policy hindrances must be removed and new policies supporting the transition established. The Bio-pathways Project established the Bioeconomy Council, which includes all associations interested in bio and green production. The objective of the Council is to develop a policy framework for the transition to green and sustainable production. The Council involves associations representing, for example, the forestry, agriculture, biotechnology, chemical, energy, transport, mining, pharmaceutical, and automotive industries. With a policy framework supported by so many sectors, the Council has a strong bargaining position in negotiations with government.

The Bio-pathways concept was presented at the Wallenberg Prize Symposium in Stockholm in 2012. After the presentation, the overall question in the audience was "Why do we not have something like this in Sweden?"

The transition processes differ greatly between Canada and Sweden. In Sweden, the transition is driven by individual forest industry companies. In Canada, the whole sector is being mobilized for the transition through frameworks such as the CBFA and the Bio-pathways and Value Pathways projects. These frameworks seek to motivate the interest groups of the forest and other sectors, as well as individual industrial companies, to advance toward transition.

There are logical reasons for the two different paths. The Swedish companies are larger than the Canadian ones, are much more internationally oriented (Canadian companies are interested mainly in the US and recently the Chinese markets), and financially stronger.

In seminars held in Canada, critical thinkers stated that forest industry transition requires more than just individual companies making their own transition efforts. The transition entails motivating a complete system to shift in a certain direction. In the end, however, it is up to individual companies to execute the transition.

Canada has taken a broader transition framework approach than Sweden has but, as discussed, the framework must broaden still further in order to deal with the systems shift. Swedish companies have started the transformation themselves without a framework and have been much more active than the individual Canadian companies. To achieve sustainable transitions and forest sectors, both paths are needed simultaneously: a solid framework for a systems shift and active individual industrial companies implementing the transitions.

References

Bonoguore, T. (2012). Half-time report. *Alternatives Journal, 38*, 1.

Canadian Boreal Initiative. (2003). *The boreal forest at risk: A progress report*. Ottawa: Canadian Boreal Initiative.

Cashore, B., & McDermott, C. (2004). *Global environmental forest policies: Canada as a constant case comparison of select forest practice regulations*. Ottawa: International Forest Resources Report.

Chronicle Journal. (2011, March 22). NAN targets boreal agreement.

CIELAP. (2011). *A green economy for Canada: Consulting with Canadians*. Ottawa: The Canadian Institute for Environmental Law and Policy.

Cullingham, C. I., et al. (2011). Mountain pine beetle host-range expansion threatens the boreal forest. *Molecular Ecology, 20*(10), 2157–2171.

Dominion. (2010, May 26). Reactions to Canadian Boreal Forest Agreement. *The Dominion*, Canada.

Duinker, P. N., et al. (2010). Protected areas and sustainable forest management: What are we talking about? *The Forestry Chronicle, 86*(2), 173–177.

Global Forest Watch. (2000). *Canada's forests at a crossroads: An assessment in the year 2000*. Washington, DC: Global Forest Watch.

Hill Times. (2011, December 5). Hard-hit forestry sector sees its future in the bio-economy. *The Hill Times*, Ottawa.

Library of Parliament. (2005). *Sustainable forest management in Canada: Clear policy – questionable practice*. Ottawa: Library of Parliament, Science and Technology Division, PRB 05-13E.

Luckert, M. K., Haley, D., & Hogberg, G. (2011). *Policies for sustainably managing Canada's forests*. Vancouver: UBC Press.

Mandel-Campbell, A. (2007). *Why Mexican's don't Drink Molson. Rescuing Canadian business from the suds of global obscurity*. Vancouver/Toronto: Douglas & McIntyre.

McDermott, C., Cashore, B., & Kanowski, P. (2010). *Global environmental forest policies*. London: Earthscan.

Mehrotra, S. N., & Kant, S. (2010). *Global competitiveness index for forest product industries*. Edmonton: Sustainable Forest Management Network.
National Forestry Database. (2011). *Wood supply background*. Ottawa: National Forestry Database, Ministry of Natural Resources Canada.
Natural Resources Canada. (1991–2011). *The state of Canada's forests: Annual reports 1991 to 2011*. Ottawa: Ministry of Natural Resources Canada.
Nature. (2010, May 18). Pact protects Canadian forests. *Nature, 465*, 279.
Nature. (2011, March 30). Canadian forest deal at risk. *Nature, 471*, 560.
Nilsson, S. (2012). *Mountain pine beetle infestation in British Columbia*. Consulting Report to Portusel, Portugal.
NRT. (2011). *Paying the price: The economic impacts of climate change for Canada*. Ottawa: National Round Table on the Environment and the Economy.
Odendahl, S. (2011, 25 June). Boreal deal putting down roots. *Listed Magazine*.
Ottosson, M. (2011). *The forest industry and the energy transition* [In Swedish]. Stockholm: The Stockholm School of Economics Institute for Research, SIR.
Paley, D. (2010, May 26). The Canadian boreal forest agreement reconsidered. *The Dominion*, Canada.
Scharmer, O. C. (2007). *Theory U: Leading from the emerging future as it emerges: The social technology of presencing*. Cambridge, MA: SoL Press.
Sierra Club of Canada. (2003). *The state of Ontario's forests: A cause for concern*. Ottawa: Sierra Club of Canada.
Suzuki, D., & McConnell, A. (2002). *The sacred balance: Rediscovering our place in nature*. Vancouver: Greystone Books.
Swift, J. (1983). *Cut & run: The assault on Canada's forests*. Toronto: Between The Lines.
Walton, A. (2011). *Provincial–local projection of the current mountain pine beetle outbreak*. Victoria: BC Forest Service.
Winnipeg Free Press. (2010, May 29). Native chiefs cut down celebrated forestry deal. Winnipeg: Winnipeg Free Press.
Working Forest. (2010, October 20). First nation leadership dismisses B.C. boreal forest meeting. *The Working Forest Newspaper*, Canada.

Chapter 10
Dilemmas in Forest Policy Development—The Swedish Forestry Model Under Pressure

Camilla Sandström and Anna Sténs

This chapter brings back the discussion to the Swedish situation and describes the forest policy dilemmas related to a transition of forest governance. The expected transition implies a shift in forest policy and practice in developed countries with a reduced "emphasis on timber production relative to the provision of environmental goods and services". The chapter describes a number of dilemmas and concludes that Swedish forestry policy has not managed to handle the gap between key stakeholders. Now this gap seems too wide to expect any joint contribution to the development of Swedish forest policy. Instead, the disagreements have resulted in putting pressure on the Swedish forestry model.

10.1 Introduction

The term "forest transitions," was coined by Alexander Mather (1992). It has been used to describe how stocks of forests can undergo transitions from, for example, net deforestation to net reforestation, in response to economic progress in society related to industrialization and urbanization. While most current research on forest transitions has focused on developing countries, the term has also recently been used to describe the transition from production to post-industrial or post-productive forest policy in developed countries. This latter form of transition implies a shift in forest policy and practice with a reduced "emphasis on timber production relative to

C. Sandström (✉)
Department of Political Science, Umeå University,
Umeå, Sweden
e-mail: camilla.sandstrom@umu.se

A. Sténs
Department of Historical, Philosophical and Religious studies,
Umeå University, Umeå, Sweden

E. Westholm et al. (eds.), *The Future Use of Nordic Forests*,
DOI 10.1007/978-3-319-14218-0_10

Photo 10.1 Forests as productive and recreational space

the provision of environmental goods and services" (Mather 2001: 250; Mather et al. 2006).

Forest policy transitions are influenced by numerous endogenous and exogenous factors and their interactions, including socioeconomic trends, political, and institutional factors, such as international norms and regulations, and also cultural changes involving shifts in values and biophysical factors, such as biodiversity loss and climate change (Mather 2001; Rudel et al. 2009; Lambin and Meyfroidt 2010). As many of these factors are beyond the control of individual governments, transition processes are often considered to be at least partly autonomous and to take place over a long period of time; this in turn is assumed to affect the scope for political steering and governance of the process (Kemp et al. 2007).

The revision of the Swedish Forestry Act (SFS 1979: 429) in 1993, which mandated changes in modes of forest governance and the balancing of environmental and industrial objectives, is often defined in terms of a paradigm shift (e.g., Schlyter and Stjernquist 2010; Nylund 2009) or as a policy transition (cf. Mather 2001). Both concepts suggest a partial shift in the role of Swedish forests from a pure industrial focus to a possible post-industrial or post-productive focus, where industrial production is no longer supreme. This policy change is often framed as being part of the current "Swedish forestry model," although the practical impact of the legislative changes has been questioned, in particular by Swedish environmental nongovernmental organizations in Sweden. However, while the concepts of paradigm shifts refer to prompt and actual change, the concept of policy transition offers an opportunity to analyze the extent to which the intentions of the policy change are actually implemented in practice. The latter concept thus fits the objective of this study, in the sense that, following the theory of policy transitions, we assume that the progress of this transition stretches over a long period of time and could potentially be severely hampered by, for example, exogenous global processes such as the "climate crisis," the "food crisis," and the "financial crisis"; these processes are expected to increase the demand for various forest products such as biomass in order to drive an energy transition from a system based around fossil fuels to one based on biofuels (e.g., Ministry for Rural Affairs 2011). This would be expected to increase economic competition relating to forest resources, which would affect other activities and ecosystem services and potentially restrict the scope for balancing environmental and industrial objectives (Beland Lindahl and Westholm 2012; Brännlund et al. 2010) a goal that, according to the environmental movement, is still to be achieved (SSNC 2011; WWF 2012a).

Given the exogenous global processes influencing the Swedish forestry sector in combination with the diverging interests of stakeholders, policymakers face two key questions: to what extent it is possible to govern forest policy transitions? And is it possible to influence the direction and speed of a transition and resolve trade-offs arising from the transition process? This case study analyzes dilemmas encountered in the governance of policy transitions (Kemp et al. 2007) in order to explore the potential for managing a long-term socio-ecological transition in the Swedish forest sector. A particular focus is put on exploring the actual goals of the current transition and on analyzing different stakeholders' perceptions of, and expectations for, these goals.

This chapter is divided into three parts. The first provides a brief introduction to the dilemmas associated with transition governance. The second describes previous forest sector transitions in Sweden, and the third outlines current transition dilemmas.

10.2 Dilemmas Associated with the Governance of Policy Transitions

The ability to adapt to change while also shaping it is closely related to problems of steering. Forest policy transitions therefore involve and create multiple challenges in governance, whether they are being actively driven as part of a government policy or occurring as a result of unplanned incremental changes. Kemp et al. (2007) identified at least six dilemmas (Table 10.1) that can potentially restrict the scope for governing transitions occurring under the influence of exogenous and endogenous drivers.

There is often significant dissent concerning goals, values, and implementation measures when complex issues are being dealt with. These goals are often defined, for analytical purposes, as dichotomous ideal types representing extreme viewpoints. Mather et al. (2006) for example define post-productivism as a fundamental re-evaluation of productivism challenging the traditional growth paradigm (cf also the definition of sustainable development in Beland Lindahl, Chap. 8, this book). Post-productivism can thus be seen as a radical shift of the forest sector from an emphasis on commodity to non-commodity outputs, that is, from maximizing production of material goods in the form of wood to broader objectives, including

Table 10.1 Dilemmas identified and solutions associated with the governance of transitions

Dilemma	Characteristics of the problem	Proposed solution
1. Ambivalent goals	Difficulties in identifying common problem and proposing relevant solutions	Continuous and iterative deliberation Problem structuring methods
2. Uncertainty about cause-effect relationships	Lack of knowledge, unpredictable systems	Flexible institutions to facilitate adaptation
3. Dispersed power of control	Sectorization, fragmentation	Joint decision making, network management
4. Political myopia	Political decision making fails to consider long-term consequences of policy decisions	Acceptance of transition
5. Determination of short-term steps, for long-term change	Lack of connections between short term action and long term change	Dual strategies of foreseeing and back casting
6. Risk of lock-in	Non-optimal solutions in the long-term perspective	Portfolio management

Adapted from Kemp et al. (2007)

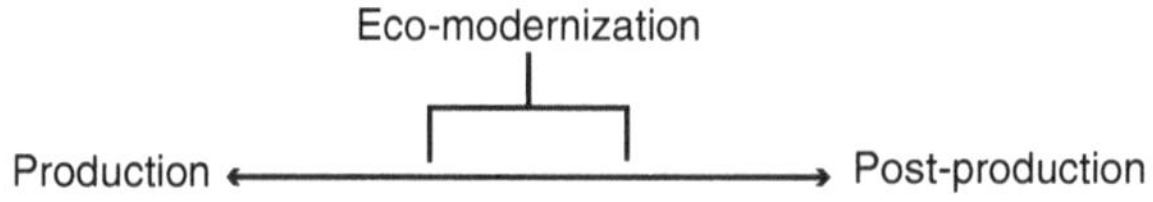

Fig. 10.1 Continuum of views on forest policies

the provision of "environmental services," such as landscape enhancement, and creation or maintenance of wild life habitats and forests for public recreation. There are, however, few empirical findings to support such a radical change in practice. This is also one of the reasons why the concept is contested and debated (e.g., Evans et al. 2002). The dichotomization fails to take into account not only the changes brought to bear by negotiations, but also the potential compromises influencing the transition process. We prefer to consider productivism and post-productivism as two axes along a continuum of views where it is possible to identify many competing visions and potential compromises (see Fig. 10.1).

Post-productivism could thus be seen as one vision connected to the goal of the ongoing policy transition: a vision that exists in parallel with other views representing less radical visions for society but where the traditional productivism and growth paradigm is retained, for example, eco-modernization. Such a vision implies that there should be a synergy between environmental protection and economic growth; that environmental problems can be solved through incremental policy development and further advancement of technology and industrialization and thus within the established principles of growth, profit, and consumerism (e.g. Baker 2006; Fisher and Freudenberg 2001, cf Beland Lindahl, Chap. 8, this book).

Differing views, together with the inevitable uncertainty regarding cause and effect in complex systems, necessitate continuous and iterative deliberation as well as the establishment of flexible institutions to facilitate adaptation. The management of complex issues spanning many levels, sectors, and time-scales due to a dispersed power of control is often subject to political myopia or "short-sightedness," which carries the risk of getting stuck with suboptimal political solutions. To overcome these dilemmas, Kemp et al. (2007) suggest a number of methods; these include trend analysis and foresight, as well as portfolio management aimed at developing both long-term ambitions and intermediary goals, which will help identify important steps and subgoals en route to addressing the main problem (Table 10.1). When analyzing the scope for governing transitions, it is therefore necessary to (i) consider the drivers affecting the transition; and (ii) identify governance dilemmas and suggest solutions to these problems to key actors.

10.3 Forest Policy Transitions in the Past

The Swedish forest sector, here broadly defined as "the economic, social and cultural contribution to life and human welfare derived from forest and forest based activities" (Beland Lindahl and Westholm 2011: 51) has undergone a number of transitions over time. The modern forest sector is often described as having evolved

Photo 10.2 Unexpected solutions may be available to complex problems

in stages, starting with the early forest management (1820–1890) stage in the nineteenth century and progressing to the sustained yield and silvicultural (1890–1945) and intensive industrial management (1945–1990) stages. The current stage (1990–2000s) has been given different names, all indicating a shift toward an ecological approach in forestry. Terms used are for example "new forestry," or the ecological, post-industrial, or post-productive stage (e.g., Öckerman 1996; Lisberg Jensen 2011; Hirt 1995; Mather et al. 2006).

The transitions were driven by science and scientists, by official forest policies, and also indirectly by social-economic processes in society at large, such as industrialization and globalization. During the first two stages, forests in more densely populated areas were used, and sometimes overexploited, for multiple purposes, such as, housing construction, fuel wood, charcoal for the iron industry. In addition, the grazing of cattle impacted heavily on tree regeneration. At the same time, timber markets opened up and forests became a valuable source of logs for timber and pulping. New entrepreneurs who were looking for short-term investments entered the forest businesses; the enterprises they funded generally did not prioritize regeneration measures to ensure the long-term sustainability of the forests. The nineteenth century has thus been characterized as an era of major ecological transition that resulted in "ecological devastation" (Östlund 1998). However, due to the introduction of more distinct tenure and access rights, the transition had social implications as well (Nylund 2009; Östlund et al. 1997). To govern and mitigate these negative effects, forest scientists worked with the government to develop a set of educational and legal institutions to exercise a level of control over the exploitation of the forests. This system was inspired by the one used in German forestry and represents the onset of rational forest management (Eliasson 2000). During the second stage in the early twentieth century, these institutions grew stronger. A number of forestry acts were passed in succession to ensure that yields of timber products remained sustainable. Additionally, regional forestry boards were established to educate and consult with private forest owners and thus ensure that the forestry acts were heeded. In addition to rational forest planning and management, there was an increased emphasis on reforestation and planting (i.e., silvicultural techniques) during this period.

After World War II, the Swedish economy peaked, and so did the market for industrial forestry. Considerable effort was invested in introducing machinery such as chain saws, airplanes, and novel silvicultural techniques such as scarification, fertilizers, pesticides, and herbicides, genetically improved seedlings, and fast-growing tree species such as the Lodgepole pine in order to achieve industrial-scale timber production. These changes were driven by the government, forest scientists, and industry in equal measure (Lindkvist et al. 2011). Although the term "multiple use" was introduced into forest policy discussions in the 1960s to indicate that the forests should be available for activities other than timber production (e.g., recreation), governmental policies continued to prioritize industrial uses of forest resources through intensive management. The level of this prioritization increased during the 1970s and 1980s, especially after the passage of the Forestry Act of 1979.

The efforts made during the twentieth century have contributed to the large growth of timber in Swedish forests that we see today. This growth has fundamentally transformed the rural landscape . It can be regarded as the result of a successfully managed policy transition aimed at ensuring high production and sustainable yield during an era when the optimistic "high modernist" mindset unified a majority of Swedish policy- and decision makers, namely, a strong faith in humanity's ability to achieve scientific and technological progress and to master nature (including human nature) so as to accomplish a rational and efficient industrial social order (Scott 1998).

However, the high modernist mindset was challenged during the postwar period. By the 1960s the public had begun to express aesthetic, recreational, and ecological concerns, which had an impact on the forest debate. New environmental institutions were established and laws were passed, such as the Nature Conservancy Act (SFS 1964: 822) and the new Swedish Environmental Protection Agency (1967), making environmental concerns a more important element in forest policies (Lisberg Jensen 2011).

10.4 From Industrial to a Post-productive or Eco-Modern Forest Policy?

As mentioned in the introduction, Sweden's current forest policy transition dates back to the early 1990s, when the Swedish government initiated a policy process to develop a forest policy that would integrate ecological considerations with modern forestry practices (Lisberg Jensen 2011).

A parliamentary committee was appointed to assess the policy and propose changes that would promote biodiversity conservation in parallel with productive forestry. Experts estimated that it would be necessary to increase the size of protected forest areas from 0.4 % of the country's total productive forested land to roughly 15 % in order to fully preserve current biodiversity (SOU 1992: 76). However, the idea of permanently setting aside large areas of productive forest did not appeal to the political majority. Instead, the preferred solution involved a combination of somewhat larger reserves (covering far less land than the 15 % that was requested) and the adoption of management practices in productive forests that would more closely mimic natural processes (e.g., preservation of some dead wood and the retention of some living trees). A third approach was also discussed, which involved zoning by separating the landscape in reserves, productive forests, and intensively managed forests. The latter preferably in locations close to industrial sites (for a modern version of this "triad approach," see Ranius and Roberge 2011). However, this alternative was not considered compatible with neither private land ownership nor the right of public access under which forest land is accessible to everyone. Intensively harvested dense forests close to industrial sites (and thus also close to urban areas) would make it harder for people to make use of their rights of access (SOU 1992: 76).

In parallel with these developments in Sweden, forest laws were being revised in many other countries as well. On a global scale, the Rio Conference (1992) adopted the Convention of Biological Diversity (CBD), and formulated a set of Forest Principles (Andersson 2007). The drivers affecting the forest policy transition were thus to a large extent global in origin (Schlyter and Stjernquist 2010). However, the outcome of the process is often defined in terms of more endogenous factors, as a compromise between key stakeholders such as the environmental movement and the forest industries. The new forest policy, which places equal weight on industrial and environmental objectives, also included an extensive deregulation of forestry operations described as "freedom under responsibility," which gave forest owners greater leeway in managing their forests in exchange for agreeing to achieve the new and more stringent environmental goals (Bush 2010). All the stakeholders consequently "had an impression of a new paradigm for forest management being introduced, concurrent with global developments towards more broadly understood sustainability and multifunctionality" (Nylund 2010: 11). The forest policy transition could thus be seen as a compromise between environmental interests, who wanted a more radical policy, and industrial interests, who wanted to focus on traditional forestry. As a consequence the policy could be framed within the concept of eco-modernization rather than as a shift from productivism to post-productivism.

10.5 A Successful Transition?

As mentioned above, transition processes that take several years to implement are often associated with a number of dilemmas such as dissent concerning fundamental goals, values and means, and uncertainty about long term effects (Table 10.1). This is obviously also the case in Sweden. The ability to handle these dilemmas and shape change in accordance with established goals is thus crucial for the overall realization of policy transitions (Kemp et al. 2007). Although the revised policy, adopted in 1993, has survived five governments (three liberal-conservative and two social democratic) with only minor changes, the Swedish forest policy transition has been characterized by ambivalence both in terms of overarching objectives and the future role of the Swedish forest. If we delve more deeply into statements made by key stakeholders in relation to recent referral processes, the compromise between environmental interests (here represented by the Swedish Society for Nature Conservation; SSNC) and the Swedish Forest Industry Federation (SFIF) seems to be particularly and increasingly fragile.

The SSNC at this time endorsed the forest policy transition and agreed that environmental conservation should be integrated into the framework of industrial forestry such that conservation becomes a natural and integral part of the business. However, the SSNC did not consider the measures aimed at protecting biodiversity in the revised forestry act to be sufficient even when they were agreed. Conversely, the SFIF was rather reluctant to adopt the eco-modernist approach and stated that it would be impossible to optimize both objectives on every hectare. According to the SFIF, it would have been more effective to allow more intensive production in parts of the landscape and to focus on nature conservation in other areas (Ministry of Agriculture 1993).

In 2006 when the forestry transition process was evaluated, the assessor concluded that the new policy generally worked well. It was, however, suggested that the portal paragraph should be amended to specify the role of the forest as a renewable resource, introduce a more specified timber production objective in parallel with the protection objectives, and also to strengthen the multifunctional role of forests as a tool for coping with climate change as well as the social and cultural aspects of forest management (SOU 2006: 81). Although this suggestion would more firmly establish the Swedish forests as a common pool resource, it was supported by the SFIF (SFIF 2007). However, the SSNC considered as incorrect the conclusion that the policy had managed to strike a balance between the production and protection objectives. Instead, they claimed that production was still constantly prioritized over protection and that the eco-modernist objective of "as far as possible, [trying] to merge an economically sustainable use of the forests with environmental and social objectives" had not yet been fulfilled (SSNC 2007).

Although the specified timber production objective was dropped and the outcome of the assessment led to a further institutionalization of multifunctional aspects in parallel with wood production (i.e., the strengthening of a number of functions such as climate aspects and social values), it did not bridge the gap between different opinions. On the contrary, the gulf in the attitudes of the key actors was further widened; the government decided to investigate the scope for intensive forest production on abandoned agricultural and forested land of low value for nature conservation (Larsson et al. 2009) in response to global trends such as "changing energy systems, emerging international climate polices, changing governance systems and shifting global land use systems" (Beland Lindahl and Westholm 2011). A research-based inquiry was established and concluded that the positive effects of intensive forestry as a tool for managing climate change would outweigh the negatives ones if it was conducted on a limited scale. It was consequently suggested that intensive forestry should be trialed over a period of 20 years within the framework of a combined strategy for adaptive forest management monitored by a steering committee (Larsson et al. 2009). While the stakeholders with a vested interest in timber cultivation were unequivocally positive toward this proposal, the SSNC and other environmental interest organizations completely rejected it on the basis that it would not fit into the current forest policy and represent a major deviation from the compromise based on eco-modernist ideals from 1993 (Lindkvist et al. 2011).

All in all, the shift in focus that began in 2006 and reemphasized forest production, growth, and even intensive forestry as a method of coping with climate change is aligned with the objectives of the SFIF, while the SSNC has been deeply skeptical of arguments that such an approach can promote forest production and growth while also stemming the tide of climate change. By referring to other scientific results the SSNC claim that the climate-based arguments are both misleading and unrealistic. While both actors still seem to support a policy approach in which ecological as well as economic values are promoted, the conflicts over how the environment is best protected are becoming stronger, pitting different methods of climate change mitigation against each other (Lindkvist et al. 2011). Judging from the recent referral rounds, the gap between key actors in terms of their perception of the goals of the transition seem to have become wider over time, undermining the already fragile compromise between the environmental movement and the forest industry.

10.6 "The Forest Kingdom"—Current Efforts to Govern Forest Policy Transition

What efforts have thus been made to bridge the gap and overcome the dissent on the fundamental goals of the Swedish forest policy? As part of the transition process, the government has decided to evaluate the system of 16 environmental quality objectives including the "sustainable forests" objective, stating that the "value of forests and forest land for biological production must be protected, at the same time as biological diversity and cultural heritage and recreational assets are safeguarded" (Swedish government 2010). The evaluation committee, which consists of members from seven parliamentary parties as well as experts appointed by the government, is responsible for developing strategies supported by a broad political consensus that encompasses milestones, policies, and measures, in particular, unclear areas that require long-term political considerations. As the mandate of the committee is valid until 2020, it could potentially provide political stability that will prevent reforms from being disrupted, and avert political myopia. The committee has also been instructed to deal with uncertainty by developing future transition paths for long-term change by using methods such as scenario analysis. The committee has presented a couple of strategies on how to deal with issues such as sustainable land use in the future.

The minister for rural affairs also emphasized the importance of Sweden's forests in stimulating growth and jobs, describing them as "our green gold" (Erlandsson 2009). This campaign which is linked to a number of hearings, and dialogue meetings has been transformed into an action plan, "The forest kingdom— with values for the world," which is centered on three key words: "sustainability, competitiveness, and accountability." The action plan, which is the first step in embodying the vision of the forest kingdom, can be seen as an attempt to create a transition platform bringing together all the relevant key actors. While the SFIF (2011) welcomed the platform and looked forward to making constructive contributions to the establishment of the forest kingdom, the SSNC expressed a completely different opinion. According to the SSNC, the role of forests in rural development and job creation cannot be based on increased extraction of wood raw material, and as the Swedish forestry model is not sustainable from a conservation perspective, any new forest-related jobs must come from the tourism industry or other industries that do not require more trees to be felled (SSNC 2012).

Based on similar arguments, the SSNC also decided to leave the "Dialogue on environmental concerns," a series of workshops with a broad participation from forestry, nature conservation, and public agencies, under the aegis of the Swedish Forest Agency. Although the purpose of the process was to improve the effectiveness of the forest policy, to bridge the gap between different actors on what aspects of the policy should be prioritized, and to assess monitoring systems that could form the basis of adaptive management processes, the SSNC perceived the agenda to be too narrow and insufficiently focused on scientific evidence relevant to the environmental quality objectives and the incorporation of the Habitats Directive's provisions on species protection (Articles 12–16) (SSNC 2012).

The longer the debate on the current Swedish forest policy runs, the clearer it becomes that the key players are at odds over both its direction and content. While the government and SFIF adhere to the endorsed orientation laid down in 1993, although with incremental changes, the SSNC and other environmental NGOs are prepared to abandon the current policy by requiring a "reboot." In the heated forest policy debate the environmental interests have adopted a more radical view over time, moving from the eco-modern compromise toward a more post-productive approach. The dissatisfaction with the current forest policy compromise is related both to what environmental interests define not only as the inability or lack of will on the part of the government to implement the policy in practice, but also to the policy per se, which they consider has successively been watered down. They consequently claim a need for a new Swedish forest policy, which would be less laissez-faire and more interventionist in order to secure the fulfillment of better defined environmental objectives (e.g., WWF 2012b).

Does this mean that Sweden is heading toward an entirely new policy transition and a post-industrial stage? So far there is no sign of that. Most influential actors, governmental as well as industrial, are not questioning the possibility of reaching the stage of eco-modernization. Quite the opposite, the idea is further being encouraged by the national as well as the European Parliament and the Organisation for Economic Co-operation and Development (OECD), promoting a total transition of society into a sustainable high-tech, knowledge based bioeconomy.

10.7 Conclusion

Despite the Swedish government having tried to deal with several of the dilemmas related to transition governance (Fig. 10.1), they have apparently not managed to handle the most fundamental one that is bridging the gap between key stakeholders through continuous and iterative deliberation. Although eco-modernization has become a broadly accepted strategy, the addition of green elements to industrial forestry has not required any radical interventions or challenged the underlying market-led economic growth policies. This might be the main reason why the SFIF (which was rather reluctant to compromise in 1993) now supports the policy, while the SSNC has become more reluctant to do so. Based on the statements made by key stakeholders, the gap between them seems to be too wide to expect any joint contribution to the development of the Swedish forest policy. Instead, their disagreements are putting pressure on the current Swedish forestry model. This will in turn affect the ability of the government to influence the direction and speed of the transition and to resolve trade-offs following the transition process.

All in all, the process of establishing industrial, intensively managed forestry in Sweden seems to have been so successful that it has become hard to disrupt (c.f. Mather 2001). Critics claim that the high modernist transition has led to a lock-in situation, which is a major failure according to Kemp et al. (2007). How true this is remains to be seen. It is, however, hard to claim that the current policy transition in

Sweden is leading to post-industrial forestry as defined by Mather (2001; Mather et. al 2006) as it might have done in other countries that are less heavily dependent on industrial forestry. The current system is better defined in terms of eco-modernism, characterized by reformism rather than radicalism, with some limited redefinition of the structure of the forest policy but no radical changes to the prevailing status quo.

References

Andersson, T. (2007). *En gemensam europeisk skogspolitik? En integrationsteoretisk studie av ett politikområde på tillväxt*. Statsvetenskapliga institutionen, Umeå Universitet. [in Swedish]

Baker, S. (2006). *Sustainable development*. New York: Routledge.

Beland Lindahl, K., & Westholm, E. (2011). Food, paper, wood, or energy? Global trends and future Swedish forest use. *Forests, 2*(1), 51–65.

Beland Lindahl, K., & Westholm, E. (2012). Future forests: Perceptions and strategies of key actors. *Scandinavian Journal of Forest Research, 27*(2), 154–163.

Brännlund, R., Lundmark, R., & Söderholm, P. (2010). *Kampen om skogen: Koka, såga, bränna eller bevara?*. Stockholm: SNS Förlag. [in Swedish]

Bush, T. (2010). Biodiversity and sectoral responsibility in the development of Swedish forestry policy, 1988–1993. *Scandinavian Journal of History, 35*(4), 471–498.

Eliasson, P. (2000). När bruk av skog blev skogsbruk. In P. Eliasson & E. Lisberg Jensen (Eds.), *Naturens nytta: från Linné till det moderna samhället*. Lund: Historiska media. [in Swedish]

Erlandsson, E. (2009, March 27). *Tal vid Lantbrukarnas Riksförbunds (LRF) stämma*. http://www.regeringen.se/sb/d/3214/nocache/true/a/125724/dictionary/true. Accessed 30 May 2012. [in Swedish]

Evans, N., Morris, C., & Winter, M. (2002). Conceptualizing agriculture: A critique of post-productivism as the new orthodoxy. *Progress in Human Geography, 26*(2), 313–332.

Fisher, D. R., & Freudenberg, W. R. (2001). Ecological modernization and its critics: Assessing the past and looking toward the future. *Society and Natural Resources, 14*, 701–709.

Hirt, P. W. (1995). *A conspiracy of optimism: Management of the national forests since World War Two*. Lincoln: University of Nebraska Press.

Kemp, R., Rotmans, J., & Loorbach, D. (2007). Assessing the Dutch energy transition policy: How does it deal with dilemmas of managing transitions? *Journal of Environmental Policy and Planning, 9*(3–4), 315–331.

Lambin, E. F., & Meyfroidt, P. (2010). Land use transitions: Socio-ecological feedback versus socio-economic change. *Land Use Policy, 27*(2), 108–118.

Larsson, S., Lundmark, T., & Ståhl, G. (2009). *Möjligheter till intensivodling av skog* (Slutrapport från regeringsuppdrag Jo 2008/1885). Umeå: SLU. [in Swedish]

Lindkvist, A., Kardell, Ö., & Nordlund, C. (2011). Intensive forestry as progress or decay? An analysis of the debate about forest fertilization in Sweden, 1960–2010. *Forests, 2*, 112–146.

Lisberg Jensen, E. (2011). Modern clear-felling: From success story to negotiated solution. In H. Antonson & U. Jansson (Eds.), *Agriculture and forestry in Sweden since 1900: Geographical and historical studies* (pp. 423–442). Stockholm: Royal Swedish Academy of Agriculture and Forestry.

Mather, A. S. (1992). The forest transition. *Area, 24*, 367–379.

Mather, A. S. (2001). Forests of consumption: Postproductivism, postmaterialism, and the postindustrial forest. *Environment and Planning C Government and Policy, 19*(2), 249–268.

Mather, A. S., Hill, G. E., & Nijnik, M. (2006). Post-productivism and rural land use: Cul de sac or challenge for theorization? *Journal of Rural Studies, 22*(4), 441–455.

Ministry for Rural Affairs. (2011). *The forest kingdom—With values for the world: Action plan*. Stockholm: Ministry for Rural Affairs. http://www.regeringen.se/sb/d/14107/a/178717

Ministry of Agriculture. (1993). *Sammanställning av remissyttrandena över 1990 års Skogspolitiska kommittés betänkanden (SOU 1992:76) Skogspolitiken inför 2000-talet och (SOU 1992:111) Den framtida skogsvårdsorganisationen*. Jordbruksdepartementet Huvudarkivet 1975–1996. E 1. Regeringsakter. Arninge: Riksarkivet. [in Swedish]

Nylund, J.-E. (2009). *Forestry legislation in Sweden*. Uppsala: SLU.

Nylund, J.-E. (2010). *Swedish forest policy since 1990: Technical report*. Uppsala: Fakulteten för skogsvetenskap>Institutionen för skogens produkter, Sveriges lantbruksuniversitet. Rapport (SLU, Institutionen för skogens produkter); 16 [Rapport]. [in Swedish]

Öckerman, A. (1996). Kalhygge eller blädning?: Svensk skogshistoria som miljöhistoria. In B. O. Linnér & J. Svidén (Eds.), *Miljöhistoria på väg: artiklar presenterade vid Miljöhistoriskt möte 1995*. Linköping: Tema vatten I miljö och samhälle. [in Swedish]

Östlund, L. (1998). Landscape change and biodiversity crisis. In A. L. Sandberg & S. Sörlin (Eds.), *Sustainability the challenge, people, power and the environment*. Montreal: Black Rose Books.

Östlund, L., Zackrisson, O., & Axelsson, A. L. (1997). The history and transformation of a Scandinavian boreal forest landscape since the 19th century. *Canadian Journal of Forest Research, 27*, 1198–1206.

Ranius, T., & Roberge, J.-M. (2011). Effects of intensified forestry on the landscape-scale extinction risk of dead wood dependent species. *Biodiversity and Conservation, 20*(13), 2867–2882.

Rudel, T. K., Defries, R., Asner, G. P., & Laurance, W. L. (2009). Changing drivers of deforestation and new opportunities for conservation. *Conservation Biology, 23*(6), 1396–1405.

Schlyter, P., & Stjernquist, I. (2010). Regulatory challenges and forest governance in Sweden. In K. Bäckstrand, J. Kahn, A. Kronsell & E. Lövbrand (Eds.), *Environmental politics and deliberative democracy. Examining the promise of new modes of governance* (1st ed., pp. 180–196). Cheltenham/Northampton: Edward Elgar Publishing Limited.

Scott, J. C. (1998). *Seeing like state: How certain schemes to improve the human condition have failed*. New Haven: Yale University Press.

SFIF. (2007, January 15). *Remiss angående Mervärdesskog (SOU 2006:81) – slutbetänkande från Skogsutredningen (N2004:12)*. http://www.skogsindustrierna.org/vi_tycker/remissvar_arkiv/remissvar_skog/aldre_remissvar_skog. Accessed 30 May 2012. [in Swedish]

SFIF. (2011). *Skogsindustrierna välkomnar regeringens handlingsplan för Skogsriket*. Press release 2011-08-09. [in Swedish]

SFS. (1964). *Swedish Nature Conservation Act* (Naturvårdslagen), p. 822.

SFS. (1979). *Swedish Forestry Act* (Skogsvårdslagen), p. 429.

SOU. (1992). *Skogspolitiken inför 2000-talet: huvudbetänkande* (p. 76). Stockholm: Allmänna förl.

SOU. (2006). *Mervärdesskog: Slutbetänkande* (p. 81). Stockholm: Fritze.

SSNC. (2007, January 19). *Remissvar på skogsutredningens betänkande "Mervärdesskog"* (SOU: 2006:81). (http://www.naturskyddsforeningen.se/om-oss/remissvar/skog/. Accessed 30 May 2012.

SSNC. (2011). *Policy: Skogen*. (http://www.naturskyddsforeningen.se/upload/Foreningsdokument/Policies/dok-skogspolicy.pdf)

SSNC. (2012, March 06). *Naturskyddsföreningens besked om fortsatt deltagande i olika arbetsgrupper inom den s.k. dialogen*. Letter to the Director-General of the Swedish Forest Agency, Monica Stridsman. (http://www.naturskyddsforeningen.se/upload/Foreningsdokument/Ovrigt/Brev_till%2520Skogsstyrelsen.pdf). Accessed 3 Sept 2012. [in Swedish]

Swedish Government. (2010). *Committee Directive. Parliamentary Commission on the basis of how environmental quality objectives can be achieved* (Dir. 2010: 74). The decision of the Cabinet meeting July 1, 2010.

WWF. (2012a). *Missade mål. Tio forskare om utmaningar i skogen*. (http://www.wwf.se/vrt-arbete/skog/1255442-vad-sger-forskarna). [in Swedish]

WWF. (2012b, May 28). *Satsa på en ny skogspolitik inom naturens gränser*. (http://www.wwf.se/press/pressrum/pressmeddelanden/1478729-satsa-p-en-ny-skogspolitik-inom-naturens-grnser). Accessed 30 May 2012. [in Swedish]

Chapter 11
Concluding Remarks: Forest Futures in the Making

Erik Westholm

The various chapters of this book provide a rather consistent view, pointing towards a scarcity crisis for land-based resources in a 2050 perspective. These outlooks seem to be highly linked to inefficient decision-making systems and institutional shortcomings. Without robust policy interventions from public institutions, market-driven governance systems can deliver neither sustainable production and consumption nor fairness and legitimacy at the global, national, and local levels. The long-term trade-offs require a formal legitimacy. These trade-offs between various demands on the land are key also in other Nordic countries where there are many demands on forested land.

11.1 Critical Demand Growth and Pressures on Land-Based Resources

Land use is the crucial link between human and economic activities and nature. Demographic changes are slow and stable and give us essential information on social and economic variables, even decades into the future. The demographic analysis applied in Chap. 3 is based on historical links between size of population, age structure, and social and economic development; this provides a convincing explanation as to why we should expect income growth to be a major driver of critically increased global demand for forest products in a 2050 perspective. Forest resources may become increasingly important, while at the same time coming under pressure from other forms of land use. As markets for forest products are quickly becoming

E. Westholm (✉)
Department of Urban and Rural Development,
Swedish University of Agricultural Sciences (SLU),
Uppsala, Sweden
e-mail: erik.westholm@slu.se

E. Westholm et al. (eds.), *The Future Use of Nordic Forests*,
DOI 10.1007/978-3-319-14218-0_11

Photo 11.1 Winter over the boreal forest

linked internationally, these pressures may be distributed to all forest-producing countries, both as challenges and as opportunities.

Forest futures are closely linked to the future use of land in general. The future demand for energy and food and the changing land use patterns resulting from expected climate change are all essential parts of the forest equation. These trends are likely to affect land use patterns, land availability, land prices and deforestation rates; and although these will interact with each other in complex ways, the actual details are hard to foresee. Our analysis of land use change in general, with specific reference to the fields of energy consumption and climate politics, shows convincingly why growth in global demand should be expected for both land and forest resources.

The chapter on future land use competition provides arguments for an estimated global demand increase of 50 % for industrial wood until 2030. Nearly all of this growth in demand will take place in "emerging" economies, while demand is expected to be stagnant or negative in the western economies. To be interpreted correctly, the implications of this increase need to be linked to other land use claims. The future global demand for energy has been analyzed in many scenarios over time. The scenarios are rather consistent in recognizing that demand is likely to increase until 2050 at a pace that points to a supply/demand crisis; some kind of transition toward energy efficiency and renewable sources of energy seems to be inevitable. However, while this seems to be possible from a resource point of view, questions are being raised concerning the capacity of existing institutions to provide the incentives, especially within the necessary time frame.

For agricultural production the outlooks are similar. Production must increase by 70–100 % to feed the world population by 2050. This means that an estimated additional 200 million ha of arable land are needed; this is expected to come partly from deforestation, causing increased competition between food and forests. The effective implementation of agricultural policies will be crucial in terms of achieving higher production, as will be the time perspective.

The climate issue adds to the complexity. The world has already locked itself into an inevitable increase of more than 2 °C in average global surface temperature in 2100. World Bank scenarios indicate a current course toward 4 °C. Chapters 4 and 7 elaborate on the effects of temperature increase and climate politics. The climate impacts are considered large enough to threaten long-term food security. From a land use perspective the most important climate change impact will be on agricultural production in low latitude countries in the form of decreased productivity, lost crops due to drought and flooding, and increased pest and disease outbreaks, which in turn may lead to land being cleared. Finally, a forest-related conclusion is that despite uncertainties concerning international climate commitments under the United Nations Framework Convention on Climate Change, voluntary markets for forest-based offsets are flourishing. However, if carbon is to be treated as a product, payments for offsetting the generation of carbon will need to be based on performance. An efficient carbon market will thus require proper carbon accounting and verification systems. Whether such systems will evolve remains to be seen.

The trends discussed so far on forestry, agriculture, energy and climate change, could lead to extensive land use changes. Effects may be direct or involve displacements of land use with rebound, cascading, or remittance effects. The most advanced land use studies aim to integrate several physical, economic, and social parameters in order to assess their integrated impacts, including indirect land use changes. A general conclusion is that the various processes must be assessed together as they are closely related; for instance, increased food production is dependent on efficient climate mitigation which, in turn, is related to the future use of energy.

11.2 The Absence of Global Governance

Altogether, the various scenarios and examinations presented are quite consistent in pointing toward a scarcity crisis for land-based resources in a 2050 scenario. We have found it essential to examine the possible policy response to these expected resource tensions between supply and demand, between resources and needs, and between short-term needs and long- term sustainability. The global challenges are indeed related to globalization itself and have to be understood primarily as governance problems. The increased complexity of actors and interests in multilevel and polycentric policymaking processes is a recurring theme, as elaborated in Chap. 6. Policy decisions concerning the future global food system will have decisive effects on future hunger and the amount of food produced, and it will also affect other types of land use, including forests. Yet, it is still the nation states which are key to ensuring that agreements are transformed into actions. Such arrangements are made in negotiations between nation states, and the commitments made are based for the most part on national contexts and national data. National governments are essential for the implementation of politics, as parliaments must transform "global" agreements into national laws and regulations. Moreover, national institutions play a key role in monitoring and supervising the agreements.

The pessimistic outlooks related to future land scarcity are based on data from futures studies and scenario projects. They seem to be linked, not so much to the lack of natural resources, but to inefficient decision-making systems, institutional shortcomings, and the fact that transitions toward sustainable production and consumption patterns are not taking place in accordance with recognized needs. With efficient governance systems, our study found, the world could have enough productive forest and agricultural land to meet the demand for food, wood products, and bioenergy in 2050. It is also clear from our analysis that without strong interventions from government institutions, market-driven governance systems can deliver neither the necessary changes toward sustainable production and consumption nor fairness and legitimacy at the global, national, and local levels. The long-term trade-offs needed require a formal legitimacy. Many of the chapters put across the urgent need for new forms of global governance without which there is a risk that the steps required will not be taken in time. Chapter 7, examining the steps toward global climate mitigation, provides a detailed view of the problems of establishing effective global agreements.

Research carried out in the Future Forests program reveals that there are two main strategies among forest sector actors with respect to futures scenarios. One school of thought holds that resource shortages can and will be solved by technological solutions and measures to enhance efficiency in consumption and production. Related to this perspective is the position that climate change is either not a pressing problem or that it is a problem that can be solved by technological progress. The other school of thought focuses on expected long-term scarcity as a major threat that can be resolved only by radical transformations of production and consumption systems. The scarcity may result from an actual lack of natural resources or from expectations that these resources cannot be efficiently mobilized. Related to this view are concerns that climate change policies may not be able to meet these challenges. It could also be argued that the resource crisis will lead to radical change, dominated either by resource conflicts and increasing welfare gaps or as the result of governance/agreements/policies that could lead to long-term resource-efficient production and consumption.

The two schools can be traced to dialogs on the future that have taken place for many decades in western economies. Actors in both camps use environmental arguments, for example, climate change mitigation strategies, to support and legitimize their strategies. In this situation, sustainable forest management is contested ground, as national governments seem to have shrinking powers to protect their forests from commercial exploitation at the expense of common-good values.

What are the governance implications for the Nordic forestry of the expected global pressures? The Nordic forest sector is now going through a "second generation" of globalization. The twentieth century brought an industrial wood regime characterized by expansive mass production across the northern forests. Nordic forest resources were exported to all major western economies. In this second phase, production is moving to the south, to low-cost countries and regions with rapid forest growth. The future of the Nordic forest companies will be less tied to their boreal origin. The projected global long-term growth in demand for both land and forest resources offers both challenges and opportunities for the future use of forests in the Nordic countries. The forest industries located in boreal regions will be subject to further competition from abroad and especially from the south, not least in their already declining European markets. This competition is likely to be intense in the pulp and the energy segments.

11.3 Scaling the Future: The Essential Role of Time Horizons

A careful assessment of time horizons is key to understanding the complex nature of the challenges that boreal forestry is facing. Chapter 4 on land use and Chap. 5 on bioenergy futures reveal that the challenges and opportunities appear to be fundamentally different, depending on which time period we focus on. First, in a 10–20 year perspective, the increased competition over supply in global forest products markets appears to be a dominating characteristic. Huge investments are being made in forest plantations in the south, where the competitive advantages are substantial both in terms of the natural conditions and of the social and economic factors related to low wages and infrastructural investment costs. For instance, the eucalyptus plantations in Asia and in Latin America offer substantial a new and relatively cheap supply of pulp. In a longer 2050 perspective, the critical growth in demand foreseen in several of this book's chapters, may increase competition between forestry and other forms of land use; however, it may also offer opportunities to increase the prices of forest products, which could motivate intensification of forest production.

All these expectations and calculations about the availability of natural resources in a distant future are beset by huge uncertainty. For instance, future demand for pulp and printing paper will depend on how newspaper reading habits develop among hundreds of millions of people in the BRIC countries. The GDP growth taking place in those countries could indicate a rapid growth in paper consumption; but it is also accompanied by a technological "switch" that could make printing paper an outdated "technology." While the eucalyptus plantations in Latin America will offer a substantial new supply of relatively cheap pulp, investments are being made under the risk of failing expectations on growing global markets. This adds to the uncertainty regarding boreal forestry. The implications of increased competition from a greater supply in the south should be assessed in conjunction with risky market outlooks due to technological transformations. The further progress of the climate problem is a wild card adding to uncertainty in all natural resource-based activities.

11.4 The Future Use of Nordic Forests

The Nordic forestry model with its tradition of consensus between public and corporate interests in the forests has frequently been associated with efficient and sustainable forest management. Its existence has often been attributed to the compromise between environmental and commercial interests made 20 years ago. However, during the last few years the idea of the Nordic forest model has been increasingly called into question, especially given the sustained conflicts between environmental and commercial interests. In Sweden, the model has clearly failed to maintain and produce biodiversity to meet Swedish national environmental objectives. Chapter 10 concludes that the Nordic forestry model, which was put forward as a solution to various dilemmas of forest management, has limitations when it comes to actual land-based activities in the forests. The model has not offered the necessary basis for bridging the gaps between stakeholders, with the state failing to appropriately address the complex field of forest policy.

So what is the outlook for the next decades? Conflicting interests in relation to forested land are more likely to sharpen than to disappear. The forests are, perhaps more than ever, expected to deliver a broad range of products and ecosystem services such as recreation, biodiversity protection, and other subsistence uses. Interest groups outside the forest sector may intervene more, and conventional forest sector governance and planning may become obsolete. In the long run, the expected increase in demand and higher prices for bioenergy feedstock and sawn timber will drive forest management toward intensification, even in the boreal regions.

Conflicts between the use of forest land for producing timber, fuel, food, and a range of other ecosystem services, including those related to climate change, are becoming increasingly apparent. Chapter 8 on key actors' visions and strategies provides important input to the crucial question of how forest politics in general will be transformed and how the forest sector will position itself in that transformation. Which actors will remain in the traditional industrial camp? Which will seek new roles and alliances that can help govern the transition from a traditional industrial/productivist growth paradigm to a system based on post-industrial values, where the forest sector can be seen as a part of the solution to the expected climate and energy crisis?

The developments expected call for a forest policy renewal. In the introduction to the book, the term *forest transition* was used to describe the transition from production to post-industrial or post-productive forests in developed countries. The Nordic countries have taken the first steps in reversing forest decline by being able to combine industrial forestry with an expanding forest cover. Forest transition, however, also implies a shift in forest policy so as to place forestry in the center of a green economy. In practice it means reduced emphasis on bulk production and an upgrading of the provision of environmental goods and service. It may include endorsement of the relative advantages of forest products from a climate mitigation perspective. Chapter 9 offers an interesting outline of the recent transformative process in the Canadian forest sector. Perhaps because Canada had never experienced anything like the Nordic Forest model, the crisis in Canadian forestry developed into a major multi-actor transformative process. The confrontations between environmental non-governmental organizations (E-NGOs), the government, and the forest industry reached the point where it was in their mutual interest to produce a national agreement on the future of the sector. The result was an agreement much broader than the idea of the Nordic forestry model. It involved not only principles for forest management but also the transition of the forest sector toward "greener" bio-based industrial production: an association of actors interested in green and sustainable production. This was also a reorientation to meet the increasingly greener consumer demands on forest products. The Bio-pathways project, however, has seen examples of both successful implementation and sustained disagreements over forest management. Obviously, parties with a long record of disagreement over how to manage the forests and make use of the values produced in them take a long time to establish mutual trust.

Can relevant stakeholders from the forest sector, civil society, and state in the Nordic countries work together to establish a forest transition that addresses the trade-offs necessary for an agreement on sustainable forest management? This is discussed in Chap. 10 on dilemmas facing the Swedish forestry model as a problem related to the state's failure to date to bridge the gap between various interests.

Perhaps this is an effect of the dominant role of the Nordic forest companies in forest politics, which has allowed them to implement the forest model according to their own interpretation. The lack of trust is not limited to the E-NGOs; it is also widely criticized in public debate on the current situation in Swedish forests. It may lessen the chances of the forest sector taking the lead in the development of a green economy.

11.5 Forest Futures—Uncertainty and Choice

We have stated that the idea of futures studies has roots in the forest sector and is a normal part of forest management. Not least in the boreal areas, where forest growth is slow with turn-around cycles of 80–100 years, it is necessary to engage with the long term; investments in forests are bound to rely on estimates, forecasting, and scenarios for many decades into the future. Nonetheless, during the forest production cycle, vast changes take place in other parts of the system. Technological breakthroughs, knowledge revolutions, and value transformations are some examples of processes with much shorter cycles and with wide-ranging effects on the choices and efforts made in the forests. While working on this book we have seen weak signals of emerging changes that may, within a quite limited period of time, call into question much of our understanding of the challenges and options facing Nordic forestry. New findings regarding fossil energy and more dramatic climate scenarios are only two examples. However, we must also be aware that many things change slowly and that change always takes place in relation to continuity. The chapters of this book all, in one way or another, elaborates on the relation between the changing and the static. Understanding these relations helps us discuss the options for the future, the choices to be made, and the different ways we can govern our future. All this, however, requires the future use of forests to be treated as subject matter for all citizens to discuss.

Index

E. Westholm et al. (eds.), *The Future Use of Nordic Forests*,
DOI 10.1007/978-3-319-14218-0

E

F

G

I

L

M

N

MIX
Papier aus verantwortungsvollen Quellen
Paper from responsible sources
FSC® C105338

If you have any concerns about our products,
you can contact us on
ProductSafety@springernature.com

In case Publisher is established outside the EU,
the EU authorized representative is:
Springer Nature Customer Service Center GmbH
Europaplatz 3, 69115 Heidelberg, Germany

Printed by Libri Plureos GmbH
in Hamburg, Germany